Dinghafte – "thingliness" p. 75

Finding Time

Cover vignette of Monhonk mountain silhouette courtesy of the Publications Office, State University College, New Paltz, New York.

Studies in Modern German Literature

Peter D. G. Brown
General Editor

Vol. 61

PETER LANG
New York • Washington, D.C./Baltimore • San Francisco
Bern • Frankfurt am Main • Berlin • Vienna • Paris

Timothy Torno

Finding Time

Reading for Temporality in Hölderlin and Heidegger

PETER LANG
New York • Washington, D.C./Baltimore • San Francisco
Bern • Frankfurt am Main • Berlin • Vienna • Paris

Library of Congress Cataloging-in-Publication Data

Torno, Timothy.
Finding time: reading for temporality in Hölderlin and Heidegger / Timothy Torno.
p. cm. — (Studies in modern German literature; vol. 61)
Includes bibliographical references and index.
1. Hölderlin, Friedrich, 1770–1843—Criticism and interpretation. 2. Heidegger, Martin, 1889–1976. 3. Space and time. I. Title. II. Series.
PT2359.H2T64 831′.6—dc20 93-37804
ISBN 0-8204-2315-7
ISSN 0888-3904

Die Deutsche Bibliothek-CIP-Einheitsaufnahme

Torno, Timothy:
Finding time: reading for temporality in Hölderlin and Heidegger / Timothy Torno - New York; San Francisco; Bern; Washington, D.C./Baltimore; Frankfurt am Main; Berlin; Wien; Paris: Lang.
(Studies in modern German literature; Vol. 61)
ISBN 0-8204-2315-7
NE: GT

The paper in this book meets the guidelines for permanence and durability of the Committee on Production Guidelines for Book Longevity of the Council on Library Resources.

Printed in the United States of America.

Contents

Table of Abbreviations

A *"Hölderlins Hymne 'Andenken'"*

B&T *Being and Time*. Trans. John Macquarrie and Edward Robinson. New York: Harper & Row, 1962

GSA *Hölderlin. Sämtliche Werke* (called the "Große Stuttgarter Ausgabe"), ed. Friedrich Beißner

HH "HÖLDERLINS HYMNE: *'Wie wenn am Feiertage...'* " in *Erläuterungen zu Hölderlins Dichtung*

HWD *"Hölderlin und das Wesen der Dichtung"* in *Erläuterungen zu Hölderlins Dichtung*

PLT *Poetry, Language, Thought*, includes *"The Origin of the Work of Art,"* translated by Albert Hofstadter

SuZ *Sein und Zeit*. 16th ed. Tübingen: Max Niemeyer, 1927

UK *"Der Ursprung des Kunstwerkes"* in *Holzwege*

[*Grosse Stuttgarter Ausgabe*, II,1, 118–120]

Wie wenn am Feiertage...

(St.I)

Wie wenn am Feiertage, das Feld zu sehn,
Ein Landmann geht, des Morgens, wenn
Aus heißer Nacht die kühlenden Blize fielen
Die ganze Zeit und fern noch tönet der Donner,
In sein Gestade wieder tritt der Strom,
Und frisch der Boden grünt
Und von des Himmels erfreuendem Reegen
Der Weinstok trauft und glänzend
In stiller Sonne stehn die Bäume des Haines:

(St.II)

So stehn sie unter günstiger Witterung
Sie die kein Meister allein, die wunderbar
Allgegenwärtig erzieht in leichtem Umfangen
Die mächtige, die göttlichschöne Natur.
Drum wenn zu schlafen sie scheint zu Zeiten des Jahrs
Am Himmel oder unter den Pflanzen oder den Völkern
So trauert der Dichter Angesicht auch,
Sie scheinen allein zu seyn, doch ahnen sie immer.
Denn ahnend ruhet sie selbst auch.

(St.III)

Jezt aber tagts! Ich harrt und sah es kommen,
Und was ich sah, das Heilige sei mein Wort.
Denn sie, sie selbst, die älter denn die Zeiten
Und über die Götter des Abends und Orients ist,
Die Natur ist jezt mit Waffenklang erwacht,
Und hoch vom Aether bis zum Abgrund nieder
Nach vestem Geseze, wie einst, aus heiligem Chaos gezeugt,
Fühlt neu die Begeisterung sich,
Die Allerschaffende wieder.

(St.IV)

Und wie im Aug' ein Feuer dem Manne glänzt,
Wenn hohes er entwarf; so ist
Von neuem an den Zeichen, den Thaten der Welt jezt
Ein Feuer angezündet in Seelen der Dichter.
Und was zuvor geschah, doch kaum gefühlt,
Ist offenbar erst jezt,
Und die uns lächelnd den Aker gebauet,
In Knechtsgestalt, sie sind erkannt,
Die Allebendigen, die Kräfte der Götter.

(St.V) Erfrägst du sie? im Liede wehet ihr Geist
Wenn es der Sonne des Tags und warmer Erd
Entwächst, und Wettern, die in der Luft, und andern
Die vorbereiteter in Tiefen der Zeit,
Und deutungsvoller, und vernehmlicher uns
Hinwandeln zwischen Himmel und Erd und unter den Völkern.
Des gemeinsamen Geistes Gedanken sind,
Still endend in der Seele des Dichters,

(St.VI) Daß schnellbetroffen sie, Unendlichem
Bekannt seit langer Zeit, von Erinnerung
Erbebt, und ihr, von heiligem Stral entzündet,
Die Frucht in Liebe geboren, der Götter und Menschen Werk
Der Gesang, damit er beiden zeuge, glükt.
So fiel, wie Dichter sagen, da sie sichtbar
Den Gott zu sehen begehrte, sein Bliz auf Semeles Haus
Und die göttlichgetroffene gebahr,
Die Frucht des Gewitters, den heiligen Bacchus.

(St.VII) Und daher trinken himmlisches Feuer jezt
Die Erdensöhne ohne Gefahr.
Doch uns gebührt es, unter Gottes Gewittern,
Ihr Dichter! mit entblößtem Haupte zu stehen
Des Vaters Stral, ihn selbst, mit eigner Hand
Zu fassen und dem Volk ins Lied
Gehüllt die himmlische Gaabe zu reichen.
Denn sind nur reinen Herzens,
Wie Kinder, wir, sind schuldlos unsere Hände,

(St.VIII) Des Vaters Stral, der reine versengt es nicht
Und tieferschüttert, die Leiden des Stärkeren
Mitleidend, bleibt in den hochherstürzenden Stürmen
Des Gottes, wenn er nahet, das Herz doch fest.
Doch weh mir! wenn von

Weh mir!

(St.IX) Und sag ich gleich,

Ich sei genaht, die Himmlischen zu schauen,
Sie selbst, sie werfen mich tief unter die Lebenden
Den falschen Priester, ins Dunkel, daß ich
Das warnende Lied den Gelehrigen singe.
Dort

Acknowledgments

I would like to express my gratitude for the extraordinary helpfulness and patience of the Hölderlin Archiv, housed in the Landesbibliothek, Stuttgart, and in particular of Dr. Werner Sohnle and Frau Marianne Schütz, without whose assistance during my stay this work could never have been completed.

To Prof. Winfried Kudszus, for the many hours of consultation and encouragement, I extend my heartfelt thanks.

To Peg Hickox, for her long-suffering tolerance during the process of research, writing, and submission of this work, my admiration and gratitude. Few indeed would have been so understanding.

Finding Time:

Reading for Temporality in Hölderlin and Heidegger

Chapter One

The Problematics of Literary Temporality

ANYONE SETTING OUT TO STUDY "time" or "temporality" in a literary text has, in a sense, already arrived too late. There can be no suggestion of unprejudiced, "naïve," viewing of the text, since it has already been read—indeed, it may be the text's own familiarity that frees the reader to consider its temporality as an issue. Unlike so many subjects of literary inquiry, temporality cannot "stand still" long enough to be "fixed" in any stable place within the discourse. Awareness of its importance comes only after it has already been at work; in this way its self-presentation initially takes the form of an absence. Consequently, the first task facing a presumptive investigation of temporality will be *methodological*. What kind of methodology offers itself as plausibly appropriate to eliciting the working of temporality in literature?

All texts, taken as linguistic constructs, contain words with temporal meaning, such as verbs or temporal adjectives. Additionally, many words obviously refer to temporal phenomena, such as "morning," "spring," "birth," etc. A linguistic or structural approach to the study of temporality could begin by charting the usage of such words within the text. Assuming that no text, particularly a "literary" text, arranges itself completely randomly with respect to the words it uses, the STRUCTURE of the text would seem to indicate something about its message. For example, the opening and closing lines of a poem are often crucial to its understanding. A theme which appears prominently in those areas would provide a clue to understanding the poem as a whole, though subject to modification by the other parts. A text speaking predominantly in the past tense creates surprise if it shifts to the present. Repetition creates expectation. Using such simple means, and others not quite so simple, literary texts are able to discourse with temporality and on temporality.

A structuralist approach seeks meaning through the distribution and use of words in the text. But not all words are equal: *"the French Revolution,"* for example, or *"Rome"* (a particular city), locate the text within a historical discourse with its own temporal claims. History brings the free play of structural differences down to earth, which is in every case either "ours" or "not-ours." Since a particular text is always in some way situated in history, with which it identifies or against which it revolts, and since any possible reader will also be historically situated, a study of temporality in the text will have to take account of the temporal relationship between the text and the historical tradition out of which it has emerged. Further, philology has shown that literary forms themselves are subject to historical change, thus providing yet another basis for temporal reference. While some literary forms, like myth or folktale, may appear free from all but the most trivial historical reference, most texts will be seen to take part in a play of reference with their historical time or their antecedents.

But neither structural nor historical investigations go very far towards understanding literary temporality. Rather, they take up unreflectingly the commonsense notion of temporality without considering their own situatedness within time. That is, each method also participates in a kind of temporality. Time is thus not merely the object of inquiry but also plays a role in the method employed to find it. The complex interrelatedness of time and text requires a detour as the most direct route to its goal. As remarked above, the time of the text has already absconded when it is first noticed. The temporality which the reader shares with the literary text takes place prior to its understanding. In order to understand its time, the text must first be read through; only with subsequent reflection can the character of its temporality be conceptualized. Such a backward-looking procedure seems to involve a *hermeneutic circle*, in which text and reader are considered as a single unit, or rather process, shifting between the whole and individual parts.

All texts, as linguistic productions, make use of temporality. Many, if not most, bring forth or display their temporality openly and consistently. For instance, a stable temporal present may be set up and maintained with only a few clear transitions. In such a case, "now" always refers to the time of writing or reading, and references to the future or past are grouped consistently around it. The historical context may appear transparent, such as in much "classic" (*not* "classical") literature: the textual present appears separate from but accessible to the moment of reading (performance). Even though Goethe's *Die Leiden des jungen Werthers* was written in 1774, its thematics of hopeless love, of *Weltfremdheit*, and of nature, continues to address readers, despite a separation of more than two centuries. Thus a text may speak to issues "across the ages." Conversely, the text may lack any stable textual present, (as in much Romantic literature). Time may then appear as a disconnected sequence of "nows."[1]

Not all texts, however, make matters so easy. Over the course of literary history certain works have acquired the reputation of being "difficult." Even after being read once, twice, or many times, they still refuse to yield up their mystery or obscurity. Yet their difficulty is not just a product of shoddy writing, which would simply be unfortunate but ultimately not interesting. These works continue to draw us back, indeed they seem to attract us even more powerfully with repeated reading, in a seductive dance of text and reader. Often their temporality is unstable or fragmentary. Rather than allowing temporality to operate unseen in its usual self-evidence or "naturalness," such texts call attention to their temporality by depriving it of stability and reliability. In my opinion, it is only through an appreciation of their temporality that they can be read at all.

Within German literature, the texts of Friedrich Hölderlin (1770–1843) have long occupied a salient position as "difficult," "inaccessible," "dark," etc. Nevertheless, successive generations of readers have registered the powerful attraction his poetry exudes. Wolfgang Binder, for example, expresses it in this way:

> Man empfand und empfindet, mit einem Wort, die Verbindlichkeit seiner [Hölderlins] dichterischen Rede, obwohl diese keine vorhandenen Wirklichkeit wiedergibt.[2]

To be sure, much of Hölderlin's reception has been tinged by biographical information (rejection in family, love, work; lack of acceptance for his writing, etc.; eventual insanity and *Umnachtung*). But, having encountered poems like *"Patmos," "Der Rhein,"* or—the focus of this study—*"Wie wenn am Feiertage...,"* no one will deny that they are a challenge to understand or interpret. We can with justice call Hölderlin's poetry "demanding" (*anspruchsvoll*): it *demands* our attentive hearing, even as it recedes from our comprehensive grasp. Just as obviously, its demand is apparently heard and responded to: its popularity—as measured by its appearance in anthologies, courses, articles—grows stronger with every year. What makes this poetry so inaccessible and yet attractive?

It is doubtful that a satisfactory answer to that question could be given solely by the discipline of literary science (*Literaturwissenschaft*). Many "reasons" have been offered for this peculiar situation, but all carry the faint odor of "rationalization," of seeking a "cover" story for what the poem has uncovered. They do not bring the poem closer; they merely encapsulate it. In speaking of *"Wie wenn am Feiertage...,"* for example, there is a temptation to account for its idiosyncrasy by reference to theoretical constructs, whether historical, linguistic, or poetological. To say, for example, that its difficulty *"inheres in the intricate, often confusing structure of its dialectic,"*[3] is to insert the poem into a discourse (*dialectic*) which soon comes to supplant it. Textual studies, elucidation of historical and literary antecedents, and even discussion within the horizon of *Geistesgeschichte*, while useful and interesting, make the text a starting point rather than a goal. Such endeavors begin with the text but soon forget it in preoccupation with its circumstances of production and reception.

Hölderlin's earlier poetry, up to mid–1799, had achieved a certain degree of attention and acceptance from the literary public.[4] Several poems had been published in literary journals and had been favorably reviewed. Leading figures, like Schlegel, Schiller, and even Goethe (temporarily), found promise in his work. But towards the end of 1799 Hölderlin began to experiment with poems in a Pindaric mode: longer, more "objective," but also more complex than the earlier lyric. These became known (through Friedrich Beißner) as the *Vaterländische Gesänge*. The first attempt at such a poem was *"Wie wenn am Feiertage...,"* which, however, remained uncompleted. It was not published until 1910, when Norbert von Hellingrath included it in his edition of Hölderlin's complete works.[5] The manuscripts of the poem were in extremely rough form, with many emendations, and with only sketches for an ending extant. A prose draft for the poem also existed, which differed in significant detail from the metric version. Ever since its publication there has been much discussion about the proper form of the poem, whether to consider it completed or fragmented, how to make use of the prose draft, and even when precisely Hölderlin stopped working on it.

None of this would seem to recommend this work to future generations of readers, but the poem itself has a significance that largely transcends the miserable state of its physical text. In this work the poet himself steps forward and speaks about the origin of his craft, the poet's special position between gods and men, and the dangers that inhere in that role. Further, the poem as a whole seems to announce the breaking of a new "time" which promises great things to come for poets and for all beings. The biographical and historical references are as intriguing as they are elusive. As a poetic treatment of time, *"Wie wenn am Feiertage..."* is significant not only in its thematics but also in the transformation of Hölderlin's style which first appears in it and which will continue to characterize all of his major work up to the time of his *Umnachtung* (1805–6).

"Wie wenn am Feiertage..." can be approached as a literary "object," having a stable existence, identity, and time frame. One could talk about "its" view of time and of its own time. Or, with a slight biographical detour, it could be seen in the light of the French Revolution and of Hölderlin's views towards it. Thus the *Feier* of *Feiertag* would be compared with contemporaneous *célébrations nationales*, as Karlheinz Stierle has done. It could be treated philologically or stylistically, focusing on the implications of its style, structure, vocabulary, genre, etc. Perhaps the greatest such excursion was by Peter Szondi.[6]

All of those studies appear useful and valuable, yet they do not go to the heart of the matter. The attractiveness and difficulty of *"Wie wenn am Feiertage..."* will not be elicited through its object-nature. By treating the poetic text as a stable object, these studies have ignored the temporality inherent in their own methodology.

Others, like George and Gundolf, have seen in Hölderlin a kind of prophet and *"der Verkünder der nahen Zukunft Deutschlands."*[7] Martin

Heidegger proclaimed him the quintessential poet (*"der Dichter des Dichters"*). Setting aside concerns with philological and philosophical standards of discussion, Heidegger claimed Hölderlin was able to bring the "essence of poetry" into speech:

> . . . weil Hölderlins Dichtung von der dichterischen Bestimmung getragen ist, das Wesen der Dichtung eigens zu dichten. Hölderlin ist uns in einem ausgezeichneten Sinne *der Dichter des Dichters*. [8]

Yet Heidegger's assertions are unsupported and seem unverifiable, perhaps even meaningless poetizing. Heidegger may claim to say the "essential" about Hölderlin, but whether or not he does so eludes many readers.

Given, however, that we *have* felt the attraction of a poem like *"Wie wenn am Feiertage...,"* how should we approach it? How are the poem's difficulty and attractiveness to be adequately described? Our initial question has led us to reconsider our method of questioning. It is clear that we are involved in a *hermeneutic process* with the text, similar to a dialogue. Access to the text is not granted all at once, it alternately attracts and rebuffs and this lively process testifies to its temporality. As a text, it thus exists within time. Western tradition has long recognized temporality in texts through two principal routes: it can appear on the one hand as FORM, and on the other as CONTENT. Aristotle, for example, after distinguishing poetry from other arts (differing "media"), found that it could be subdivided according to how it either represented different "things" (CONTENT), or how it represented them in different "ways" (FORM).[9] Both self-presentations unfortunately lend themselves to reification and thus threaten to set up an independent existence for themselves. FORM easily usurps control and issues commands about proper access to the text (itself). That is, formalist approaches to texts require that only certain aspects of the text be taken as significant: "how" the text produces meaning, regardless of "what" it says. CONTENT, the issue at stake in the text, may dominate so completely that it comes to seem self-evident and autonomous, thus obscuring its reliance on a particular FORM. The story being told, the issue being discussed, may come to seem so natural, so important or self-evident, that the reader's distance from the text (as well as involvement in the textual process) is blurred. It is all-too-easy to withdraw from an understanding of the primordial relationship of form and content to the text into a detached "gazing" at them through the spectacles of objectivity. In spite of the popularity which literary-theoretical approaches to textuality based on form and content have shown and continue to show, we shall see in the course of this study that they cannot reach the fundamental nature of temporality. Rather, it will become clear that *the temporality of the poem is something which exists prior to any split into form and content.* To understand this, we need to take a brief look at the temporality of these two terms.

How are we aware of temporality in poetry? At the most superficial level, poetry—if it is "good" as poetry, i.e., effective—clearly speaks a language that draws us in, back to itself. It is *memorable* language, we go over it again and again, we may find it recurring in our thoughts without conscious effort. This implies that the poem seeks to overcome temporality, precisely because it refuses to fade away. *"Exegi monumentum aere perennius."* Additionally, poetry, as language, makes use of linguistic signs (tense, aspect, temporal adverbs) to indicate temporal relations. Reading the text, inasmuch as it is a PROCESS, necessarily bestows temporality (PROCESSuality) on the text. Thus, the text's FORM (structure) manifests temporality as process, in terms of sequence and repetition. In literary texts the process is both *finite* (having beginning and end) and *infinite* (inexhaustible). Out of this processuality the poetic text is able to develop its rhythmic and musical capacities. It learns to "sing" in time.

Temporality appears in another way, too: as a theme, in the poem's CONTENT. The text may self-reflexively state or question its own "time." *"Wie wenn am Feiertage...,"* for example, speaks of the *"jezt,"* which is a *"Feiertag."* Temporality, when it appears as part of the CONTENT of a poem, renders it self-reflexive. The question of temporality is therefore not a matter of indifference to the text. Rather, it forms its origin and sets the course of its development. Ultimately, it is the result of a decision about how and where the process of taking a stand *begins*. In other words, the question we are seeking to answer is: *"Where do I find the 'origin' of the artwork?"* Where does it come from? Does it come from a "where" at all?

It might seem at first that a poem does not properly have an "origin," that it is just naturally "present," like a rock, a leaf, or any other "thing." It has a reliable presence as something that can be taken from the bookshelf, for example. Such a description, however, hardly does justice to what we mean by a poetic text. Even if we should just "accidentally" happen upon it, the poem speaks to us, if at all, from within an *historical* (*viz.* temporal) transmission. It is not simply "present," contextless, ready to be discussed or analyzed at any time like a scientific discovery (and today even *that* timelessness is being questioned). Rather, it exists within a discourse of the present with the past (about the future). Hölderlin's poem *addresses* us from an historical tradition and brings that tradition into our present. It *addresses* us: i.e., it gives us an *address*, a place, *einen Ort*: *es er-örtert uns*.

In speaking of the "origin of the artwork," we are plainly moving into the territory "cleared" by Martin Heidegger. Indeed, I find in certain of his texts the most far-reaching considerations of temporality in literature—and particularly in his discussions of Hölderlin. It is specifically in terms of his method or process of reading that Heidegger is so valuable to this study. As David Wood points out, *"[Heidegger's] achievement lies as much in the way he follows his problems through as in any one of the means he adopts to solve them."*[10] But his own texts are at least as obscure as those of Hölderlin; how is he then to help us? Even Karsten Harries, who is generally favorable to Heidegger's readings, complains, *"His words retain their suggestiveness, but it is difficult to say just what they suggest."*[11]

In my view, Heidegger, in his discussions of Hölderlin, *retrieves* (*wieder-holt*) Hölderlin's inaccessibility, and thus forces attention back to the process (i.e., temporality) of his own language. He enters into a hermeneutic process with the text, eschewing insistence on a linear search or a single, unified goal. Rather, it is a question of "listening" to the text, even when it speaks through reticence. His prose is apparently a farrago of neologisms, etymological figures, oxymorons and *sententiae*. If we tried to insert it into a literary-historical discussion of "*Wie wenn am Feiertage...,*" it would severely strain credulity. Christopher Fynsk, in approaching Hölderlin through readings of Heidegger, comments on the *"violence of Heidegger's interpretation of Hölderlin."*[12] In spite of Heidegger's radical and explicit contestation of the authority of Western analytical-rational thought, most commentary on his writings obstinately seeks to understand what he says in terms of that very tradition. It takes his words as "assertions" and isolates them from their context in the process of his thought. By restricting the focus of reading to the development of a coherent "position," such criticism does violence to the very project in which Heidegger is engaged. Walter Biemel refers to this approach as criticism from the "outside," pointing out that it boils down his thought into certain positions which can then be easily refuted. But what Heidegger is thus criticized for is, from his point of view, precisely the advantage of his thought.[13]

Experience shows, however, how hard it is to depict or set out that position. The divergence of opinion among Heideggerians (and those who write on him from a "non-Heideggerian" stance) is justly famous; there can be no "schools," since so little agreement has been reached about just *what* he was saying. Could it be that such criticism has been ignoring just what is most prominent in Heidegger's texts: namely, their TEMPORAL character, which entails interpretation as a PROCESS? Heidegger seldom seems so intent on establishing a position, as much as *engaging in thought* with a text or texts.

> Die Erläuterungen gehören in das Gespräch eines Denkens mit einem Dichten, dessen geschichtliche Einzigkeit niemals literar-historisch bewiesen, in die jedoch durch das denkende Gespräch gewiesen werden kann.[14]

Heidegger here contests the ability of literary history to "prove" Hölderlin's uniqueness. And yet this is not a "position" (or counter-position) he takes up, but a by-product of his continuing dialogue with the poetry. A position, after all, by encapsulating thought, removes it from its own temporality and thereby frees it for employment in "anyone's" rhetorical scheme. *"Das Wort ist Waffe,"* as both Heidegger [HH, 58] and Hölderlin (*"Am Quell der Donau"*) knew: as a "position," it can be used to exile Heidegger to the pages of error: one determines his "position" in order to show how he *failed* to make his case, how he missed, avoided, or repressed part of the subject, etc. As "positionless," however, Heidegger exerts an insidious influence on those who invoke his name, even if (*especially* if) they seek to sweep him into the "dustbin of history." His own temporalizing (his PROCESS) infects theirs and multiplies his influence. That is, their own prose and their own approaches to texts become more open to temporality—more mobile, flexible, daring—because of the insights wrested from (poor, wrong-headed) Heidegger. The present study certainly exhibits its debt to him.

I propose, therefore, to explore temporality in *"Wie wenn am Feiertage..." phenomenologically* and *hermeneutically*. Phenomenology studies that which comes to light (*ta phainomena*) in the way that it comes to light, without postulating a "reality" behind "appearances." It begins with the immediacy of experience, including not only what appears but how it appears. Hermeneutics questions texts while allowing them to question the questioner, following a process of give-and-take between text and the reader. Together, these approaches illuminate the way temporality works in texts. The choice of this method is obviously influenced by Heidegger's own description of his method: *"The phenomenology of Dasein is a hermeneutic in the primordial signification of this word..."*[15] I make no claim that our two methods will bear any similarity to each other. My reading of Heidegger will be "literary": an attempt to use the tools of literary criticism to follow Heidegger's thought more closely than could be done by simply focusing on his terms and positions. At the same time, the literary method itself must remain open to Heidegger's own insights, which may in the final analysis elude analysis. This study therefore seeks to preserve the modesty with which Robert Mugerauer begins his study of Heidegger's thought:

> The attempt to hear and think about what Heidegger says is, at the same time, the attempt to find a way whereby those texts can speak for themselves and reveal themselves. [16]

It would perhaps be well to consider Heidegger's "way" with a text before proceeding to follow it. A central place in which Heidegger discusses his hermeneutic method explcitly is in §2 of *Being and Time*: "The Formal Structure of the Question of Being." His immediate concern there is to discover the right way of approaching the question of Being. Since this question has been "obscured" (*in Dunkel gehüllt*) through the whole history of Western philosophy, it needs first to be *"made transparent, and in an appropriate way."* Heidegger is thus employing the language of phenomenology: that which appears (*to phainomenon*) is to be investigated, both with respect to the "fact" that it appears—its factuality—and to the way in which it appears—its being-"as" something. Of course, the present opacity of the question of Being, the recognition that it has been neglected, forms part of the phenomenon to be investigated. The question of the meaning of Being strikes modern ears as empty, vain, pointless. Why waste effort on such a question?—so would most people think. Yet it is precisely the awareness that something has absconded and left a void to which this question attempts to draw attention. At a given historical time, some things seem self-evident while others appear hardly conceivable, even as questions. Something has been forgotten and needs to be sought now; it is the *questioning* of Being that Heidegger attempts to initiate, not the pretence of supplying the single and sufficient answer. He begins, then, with the discovery that Being's very self-evidence has so far allowed it to escape questioning.

Just as important as recognizing the nature of the question is formulating it in a suitable way ("[*mit*] *der angemessenen Durchsichtigkeit*"). If we are to seek the answer to a question, we must first decide what path is most likely to lead to it. There is always the danger that we could *impose* an answer on it, rather than allowing *it* to make its own answer known. The fundamental though elusive problem of methodology must be considered thoroughly before the topic can ever be broached.

> Jedes Fragen ist ein Suchen. Jedes Suchen hat sein vorgängiges Geleit aus dem Gesuchten her. Fragen ist erkennendes Suchen des Seienden in seinem Daß- und Sosein. [SuZ, 5]

> Every inquiry is a seeking. Every seeking gets guided beforehand by what is sought. Inquiry is a cognizant seeking for an entity both with regard to the fact that it is and with regard to its Being as it is. [B&T,24]

In this passage Heidegger shows how his approach will differ from other attempts to answer the question of the meaning of Being. By invoking the image of seeking (*Suchen*), he opens up the possibility of including *temporality* in the problem, since time is required to bridge the gap between the present absence or hiddenness (of the object sought) and some future or possible presence. Inquiry is like seeking, which is like traveling down a path. Obviously, someone traveling on a path "gets guided beforehand" by his destination; not just any direction will do, if one is to reach one's goal. Similarly, a search could take off in many directions but would have no guarantee—in fact it would have little likelihood—of finding what it was seeking. The very words which propose the inquiry give it an implicit order or direction. In this case, by asking after the "meaning of Being," a tacit recognition of the simple meaningfulness of the question comes to light. The bulk of *Being and Time* will be occupied with working out the consequences of that question.

Heidegger, although having framed his inquiry as a search, does not conduct his search in the typical way. He does not assume that the object of inquiry (*das Erfragte*) already exists somehow, only in a deficient state (i.e., absent, unclear, non-transparent) which must be made to yield to the incursions of the investigator. Instead, he recognizes that the path of inquiry, having sought its object, returns "home," at least provisionally. That is, the inquirer, immediately upon bringing up the question, finds himself implicated in the absence of *das Erfragte*, "that-which-is-being-sought." It must return to the inquirer and *inquire after him*. Thus, the proper attitude is one of listening, not of speaking; of waiting, not of forcing ahead. The inquiry is not like a guide, but like one who is guided.

By stipulating that this inquiry is "cognizant" (*erkennend*), Heidegger excludes all other kinds of seeking, such as looking for a pair of missing glasses, which need not entail an explicit awareness of the act of seeking. When I try to remember a word or name, all that matters is that I finally recover it (or that it "comes back" to me), not what particular process was involved in its recovery. In the phenomenological-hermeneutic approach, on the contrary, the method will be "cognizant," i.e., self-reflexive. The method itself, the way of framing the question, takes precedence over the effect or answer.

> Ein Fragen kann vollzogen werden als "Nur-so-hin-fragen" oder als explizite Fragestellung. Das Eigentümliche dieser liegt darin, daß das Fragen sich zuvor nach all den genannten konstitutiven Charakteren der Frage selbst durchsichtig wird. [SuZ,5]

> When one makes an inquiry one may do so "just casually" or one may formulate the question explicitly. The latter case is peculiar in that the in-

> quiry does not become transparent to itself until all these constitutive factors of the question have themselves become transparent. [B&T,24–5]

Heidegger's method is thus accumulative as well as reflexive. It is a process which does not reach its conclusion except as a totality, a *Gestalt*. It either "is" or "is not," and thus must await the attainment of a certain threshhold. The path to that attainment requires questioning the question, so that what is implicit may be brought to light. In that way, the question becomes *"transparent to itself."*

The present study seeks to understand "temporality in literature," whereas in *Being and Time* Heidegger was after the "meaning of Being." Given that difference in focus, however, the insight which drove Heidegger's phenomenology can also be applied to our own investigation.

> Wir wissen nicht, was "Sein" besagt. Aber schon wenn wir fragen: "was ist 'Sein'?" halten wir uns in einem Verständnis des "ist," ohne daß wir begrifflich fixieren könnten, was das "ist" bedeutet. Wir kennen nicht einmal den Horizont, aus dem her wir den Sinn fassen und fixieren sollten. Dieses durchschnittliche und vage Seinsverständnis ist ein Faktum.
>
> Dieses Seinsverständnis mag noch so sehr schwanken und verschwimmen und sich hart an der Grenze einer bloßen Wortkenntnis bewegen—diese Unbestimmtheit des je schon verfügbaren Seinsverständnisses ist selbst ein positives Phänomen. [SuZ, 5–6]

> We do not know what 'Being' means. But even if we ask, 'What is "Being"?,' we keep within an understanding of the 'is,' though we are unable to fix conceptually what that 'is' signifies. We do not even know the horizon in terms of which that meaning is to be grasped and fixed. But this vague average understanding of Being is still a Fact.
>
> However much this understanding of Being (an understanding which is already available to us) may fluctuate and grow dim, and border on mere acquaintance with a word, its very indefiniteness is itself a positive phenomenon which needs to be clarified. [B&T, 25]

Heidegger has remembered Being as a question and now reminds us of our (and his) ignorance of how to answer it. In a move which will remain a hallmark of his style, he transforms a seemingly fatal problem into an incipient solution. Ignorance is not simply the opposite of knowledge, but is itself a *kind* of knowledge, a kind which is frequently encountered in everyday life(!). Every human being, of whatever degree of intellectual attainment, has some kind of understanding of what it is to "be," to exist as a human being, even though that understanding may be limited, unstable, and indefinite. Almost certainly it will not be explicit. Even so, the *clarity* with which its vagueness stands out, once attention has been called to it, remains undeniable, a "Fact" (*Faktum*).

In the years after publication of *Being and Time,* Heidegger came to view the art work, and in particular Hölderlin's poetry, as uniquely valuable points of departure for understanding Being. *"The Origin of the Work of Art"* (*"Der Ursprung des Kunstwerkes,"* 1935/6) first set out in detail Heidegger's approach to the work of art in general and asserted the particular relevance of the linguistic work of art (*Dichtung*) to understanding the nature of art itself. *"Hölderlin and the Essence of Poetry"* (*"Hölderlin und das Wesen der Dichtung,"* 1936) brought him to a consideration of the fundamental importance of Hölderlin in the understanding of poetry, and the series of *Erläuterungen* continued it into close inspection of individual poetic texts. Since we have selected *"Wie wenn am Feiertage..."* as the text whose temporality is at issue, only the essay dealing with it (*"Hölderlins Hymne: 'Wie wenn am Feiertage...'"* 1939) will be considered. Heidegger himself offers some support for this choice by referring to *"Wie wenn am Feiertage..."* as *"die reinste Dichtung des Wesens der Dichtung."*[17]

The course of the present study will thus proceed from a "direct" or "naïve" reading of the text, to a closer (yet still immanent) reading, to a consideration of Heidegger's approach (through *"Der Ursprung des Kunstwerkes," "Hölderlin und das Wesen der Dichtung,"* and *"Hölderlins Hymne: 'Wie Wenn am Feiertage...'"*) to a concluding statement which will bring all four texts into focus.

Chapter Two

"Wie wenn am Feiertage . . .": Hymn or Warning?

An Initial Reading

THE GOAL OF THIS INVESTIGATION is the temporality of Hölderlin's poem *"Wie wenn am Feiertage..."* It would seem the most direct and promising approach would be a reading of the instances of temporal words or expressions, followed by a listing and ordering of them into a form from which general conclusions could be drawn. Such is the generally accepted method of philology or of literary scholarship in general. It would of course include the relevant historical (textual and biographical) background. However, I am not going to follow that approach here, so fruitful as it may seem, because it precisely *misses* the nature of the temporality being sought. Before explaining what method I propose to adopt, however, it might be useful after all to follow philology briefly into its blind alley and make that our provisional starting point.

This poem, like so many by Hölderlin, was never published in the poet's lifetime. Even the exact chronology of its composition is still hotly debated.[1] The primary manuscript for the latest metric version falls into the period between late 1799 and the first half of 1800. It remained in manuscript until 1910, when Stefan George and Karl

Wolfskehl brought it out in the third volume of *Deutsche Dichtung* (*Das Jahrhundert Goethes*).[2] It exists in both a prose version and a metric version. The manuscript of the metric version is one of the most tangled and difficult to decipher in Hölderlin's entire corpus.

Reading the text as presented in the Beißner edition[3] for its temporal structure brings out rather quickly the importance for the poem of the "poetic present," the "NOW" (*jezt*)[4] within the poem. Aside from explicit reference to the poetic present through use of the word *jezt*, the text throughout points to its importance by comparing the past with the present (always to the advantage of the latter). The importance of the poetic present is established not just by its multiple appearances in the text (St.III,1;[5] St.III,5; St.IV,3; St.IV,6; St.VII,1), nor by the use of tense, but also by the obvious significance of what happens during those occurrences. Even the most cursory summary cannot avoid mentioning the arrival of day in the third strophe, towards which the first two strophes seem to prepare the way. That is, the structural hint offered by the repetition (and thus emphasis) of the *jezt* is reinforced by the extraordinary importance granted to the events which transpire within it. After a description of the sad and isolated—yet hopeful, anticipatory—state of the poets during Nature's "sleep," the poet bursts forth with:

> Jezt aber tagts! Ich harrt und sah es kommen,
> Und was ich sah, das Heilige sei mein Wort. (III,1–2)

The poet thus proclaims not only the approach of a new "day" (i.e., a new age), which under normal circumstances would in itself be sufficient cause for jubilation, but identifies what he has seen approaching—what the reader has previously discovered to be Nature—with the Holy. This is truly an epiphany of enormous proportions, as befits the hymnic form and the chosen subject: the nature of the poets and the production of their song. The following strophes reinforce and elaborate this idea, even casting the poets in a heroic light, since they voluntarily expose themselves to the divine storms, for the sake of receiving inspiration and passing it on, in the milder form of song, to their fellow human beings.

In such a summary, as grossly oversimplified as it is, we may nonetheless make out rather clearly the significance of the present and the past for the poet. The poetic present is proclaimed as almost apocalyptically important: it is the *kairós*—the propitious moment or *"der erfüllte Augenblick"*[6]—in which the poet approaches the god and is struck with divine presence. Thus, looking at the text linguistically, we have brought out the importance of the present moment, the time of the poem, and have seen how the poet has invested it with a joyful celebration of the long-awaited re-appearance of the Holy.

There is a snag, though. The final lines of the poem have not yet been addressed. The text of *"Wie wenn am Feiertage..."* is a fragment. Of the projected nine strophes only seven were completed, along with segments of strophes VIII and IX. How can we therefore talk about "final" lines? In the last completed strophes and in the sketches for the incomplete strophes the poem's temporality takes a radical turn away from the certainty and jubilation expressed before. The poet, having assured us of the rightness of the poets' action and of their safety, suddenly turns around and breaks forth (twice) with, *"Weh mir!"* Gaps in the text follow, left to be filled in later, and then the poet calls himself the "false priest" who has been cast into darkness by the very gods he sought to approach. *De profundis* he sings "the warning song" to those willing to hear it.

> Und sag ich gleich,
>
> Ich sei genaht, die Himmlischen zu schauen,
> Sie selbst, sie werfen mich tief unter die Lebenden,
> Den falschen Priester, ins Dunkel, daß ich
> Das warnende Lied den Gelehrigen singe.
> Dort

The "warning song" seems to refer to this very poem, the poem being sung *"jezt"* by this poet.[7] If that is so, if the poet in the poem is calling the poem he sings a "warning song" (not a song of praise, a *Feiertagshymne*) sent from the outer depths by a "false priest," then it seems that the poem itself has very forcefully repudiated its own statements about the arrival of the *kairós* and thus about the nature of the temporality in which the poem is situated. The reader is left in a quandary: are the earlier statements about temporality true and reliable, or do they only apply to *some* poems? And if the poet who speaks of these things proclaims himself to be "false" at the end of his text, how are his earlier statements to be taken? Can his word be trusted at all? Does one word supercede another, or do they exist independently of each other? Put differently, what is the status of the words in the text?

Some scholars[8] have tried to soften this dilemma by pointing to the conditional form of the lament *Und sag ich gleich, / Ich* sei *genaht...* [emphasis added] This argument seems rather weak. Although the poet has not unequivocally accused himself of falseness, still it was he who brought up the bare possibility of such a thing. This hypothetical occurrence is also set quite emphatically in first person address, whereas throughout the rest of the poem "the poets" were referred to in the third person, or in direct address (*ihr*), or as part of a group including the present poet (*wir, uns*). The only other occurrence of the poet speaking in his own voice is in the lines from St.III, quoted above, where he proclaims the awakening of Nature and the arrival of the *kairós*. Thus, the poet's own voice appears only once as bringer of praise and once as exile.

The easy way out of this quandary would be to say that the poet was not in full possession of his craft (or his mind!); or, since the poem is obviously unfinished, that we have here merely some rough edges which would have been smoothed out later, had the poem ever been "completed." This argument overlooks the fact that we *do* have an earlier, prose draft of *"Wie wenn am Feiertage...,"* in which the very same turn is taken. Some, like Peter Szondi,[9] actually attempt to interpret the missing text on the basis of this draft. While I do not feel the earlier draft can replace the empty lines of the metric version, the existence of the same turn from light to dark (in the prose sketch) at least demonstrates that it was not a spontaneous outburst in the metric version. Further, many poems by Hölderlin show a similar reflex at their end (*Friedensfeier, Patmos*) which significantly darkens their tone. Other interpreters—as dissimilar as Martin Heidegger,[10] Karl Lachmann,[11] and Hermann Pongs,[12] for example—simply ignore the ending and insist on the earlier published version of the text by Hellingrath, in which the poem stops with the seventh strophe.[13] This is clearly self-serving. While it may evade the problem, it does not thereby eliminate it. The unfinished state of the poem in manuscript does not justify suppressing such a major and evident shift. Lachmann goes so far as to claim, on the basis of the truncated manuscript, that the poem *"stellt dar, . . . wann hymnischer Gesang* glückt."[14]

To restate the problem as I see it: If we seek the action of temporality in the poem based on the explicit statements made by the poem, we are left in an aporia. The kind of time described in the final lines of the text is, in my opinion, not reconcilable with the time announced earlier (the *Feiertag* or *kairós*). That is, the poetic present is described as a special and favorable time for poets to step forward into divine presence so that they may produce their songs. At the same time, if a poet steps forward he risks ostracism (or divine blasting[15]). The present poet, however, goes further and clearly implies that he himself is unworthy of such an approach; in fact, he has already been cast into darkness. Either "favorable" (*günstig*) means something different from our usual understanding of the word, or the poet has contradicted his earlier statements about the present: it is not "favorable" but dangerous, stormy, etc. Thus, any discussion of time based on such statements, ultimately deriving from the linguistic temporal markers (tense, aspect, temporal adjectives and adverbs) and the historical evidence, seems meaningless.

If we are to take this poem seriously, and I think anyone who has read it at all attentively would agree that it deserves to be taken seriously, then we must re-examine any presumptions about temporality and about how it is to be found in this (or any other) poetic text. That will necessitate another look at the opening moves of this investigation. The breakdown of the methodology does not eliminate the need to employ it. Rather, it serves to call attention to itself *as* itself and thus focuses on the hermeneutic stance of the investigation. It is only by following the *usage* of temporality within the poem that its *significance* will be made apparent. Since the text already shows internal fissures, an overall stable time-reference has been shattered. As a figure, however, self-contradiction is only a particular way of bringing two aspects of an issue together (i.e., as mutually exclusive or combative). Such an "unnatural" (anti-mimetic) use of language is even more helpful as a guide to the temporality of the poem than a "natural" use would be. It allows the reader to draw near to the poem and yet maintain a measure of distance from it.

The following analysis of *"Wie wenn am Feiertage..."* reads the poem both linearly and phenomenologically. The linear reading is the poem's preferred one, since the poem has chosen it. Although perhaps also a “naïve” reading, it must still be taken up at least once before it can be cast aside. At the same time, the linearity of the text cannot disguise its discontinuity, which is discovered by describing phenomenologically the key terms. Finally, it is through a linear reading that the temporality of the poetic syntax and structure also becomes evident.

The First Triad (Strophes I–III)

How does one recapture the initial impression made by a poem which since has become familiar? A “new” look at a familiar poem requires that one pretend never to have seen it before. The only possible access to the past is through speculation—i.e., memory. These remarks may serve as a warning that even finding the “beginning” of our interpretation is not unproblematic. Temporality (in the form of *das Gewesene*) already informs our reading before we begin. Nevertheless, we do begin to read the poem. It does not seem hard to read, the words are familiar, everyday, with only a few “literary” terms (e.g., *Gestade, Strom, Hain*). The first strophe seems to describe a landscape on the morning after a storm. This description then serves to form the basis of a comparison with the present state of poets, who, like the objects in the landscape, are also under the all-encompassing care of Nature. All seems peaceful and promising.

By the end of such a reading, however, the reader is probably shocked and confused. What has happened to the promise of the opening strophes? It seems as if the poets, at least one of them, were not standing in such favorable circumstances after all. To answer this question, which seems to be posed by the text itself as a whole, the reader must return to the beginning and read again, with different, wiser, *erfahrener*, eyes.

He finds himself in good company. The *Landmann* too returns to his field. The field, like this poem, is familiar to him. He “goes” every day, and today, on a *Feiertag*, he goes again. What is the significance of this going which is in reality a returning? Most commentators conclude he is concerned about the state or condition of his field after the nocturnal storm. Thus, he is relieved that all is well and that the natural world seems to be reviving. A connection would thereby be prepared for the description of the state of the poets in St.II, who there also seem to be in good shape.

But if instead we think more carefully about this man, we find that he himself has a most fundamental relationship to his field. He is not merely a farmer, or peasant, or worker; rather, he is a "man of the land" (*Landmann*) who goes to "see" his land, his field. We remember here that *Land* originally carried the meaning of *Acker* (in Gothic), then "*väterliches Erbgut*";[16] in Hölderlin's native dialect, Swabian, indeed, it has this even today as its primary meaning.[17] As so often, Hölderlin has found an everyday word which points beyond its everyday use, back to a more fundamental, indeed, ontological meaning, which at the same time finds a clear analogy in the empirical world. A *Landmann* in the sense of a farmer (*rusticus*) does indeed live from the land: often he is born and buried on it; when it is damaged, he too suffers. Previously, one often used the name of the land for the name of the people dwelling on it. The *Feld*, *his* field, is thus his *Land*; like so much that is nearest to us, it is also the least known. Although he goes to his land every day, the *Landmann* does not "see" it; instead, he sees the tree which has fallen or the fruit which must be picked. In other words, the land exists for him as a "field" of concerns. It itself withdraws, so that the concerns can step forward which give the man his identity, so that he can make a "living."

We are told that today the *Landmann* goes to "see" his field. He does not go to "work" it, i.e., to produce something by means of it. He thus attempts to approach it more originarily. Temporality is what makes this more original ("seeinger") seeing possible: it produces the *Feiertag*. This is another word taken from common parlance, even today often used periphrastically for *Sonntag* (Sunday[18]). Yet, we are given no details about this day; it does not seem important to know if this is a saint's day, a Sunday, or a pagan feast day. The lack of determiners allows the word to step forward as what it originally is, a day of celebration (*feiern*). The resonance of the *Feiertag* will occupy us throughout this poem.

The *Feiertag* is, first and foremost, a "day," i.e., it allows time to step forward by calling attention to the (temporary) presence of the sun. As creatures of light, humans have always been reliant on the sun. It makes possible the discovery of a world in which they are able to dwell as the kind of being they are. Even darkness derives its meaning from the (absent) sun. Since, however, the sun comes and goes, men must "pay attention" to it, we must know the right time, what the Greeks termed the *kairós*, in which to act and the right time to refrain from acting, to wait, to rest. "Day" thus establishes time by bringing man into relation with the sun (heaven) and earth(Nature).

We have said that day is the time for action. Since the presence of the sun enables seeing, it also presents the possibility of acting. But through the repetition of the action of the sun approaching/seeing/acting/resting/re-approaching, etc., time itself withdraws from view. "Life" continues as usual and "we" disappear into its flux. In order to prevent this loss we must step out of the flux. If man is to see that he sees, to understand the nature of his sight, which is seeing his world, i.e., his "time," he needs a time "out." Such a time out is provided by the *Feiertag*. It is a time in which ordinary activity is put down, and yet men's eyes remain open. Seeing continues, but without the distractions of every day. The world can be seen as a whole and the source of sight, the sun, steps forward. Such a seeing is an ontological act, it brings man into contact with Being. It is also and equally a return: although he has not truly been "away," he has also not truly "seen," or rather, he has only seen beings, not Being.

Further, we cannot merely decide at any arbitrary "time" to call a halt and thereby come into the nearness of Being. We do not decide at all: rather, it is "given" to us. Just as the world has been given to us, so now is the *Feiertag* given. On this day, we approach the god and give back what has been given. This is the fundamental re-enactment of the creation of the world which forms the basis of myth. Only a god can "create" time, and only a god can call man to re-view it. For this renewal to occur, a return is necessary, a return to the god himself.

Such an approach, however, is not without difficulty. How should one approach the god, who has after all kept hidden for so long? How to find the way? For this, the *Feiertag* itself provides an answer: with song. Traditionally, on the *Feiertag* the poet (or priest) leads his people back to the god with songs of praise and thanksgiving.

These remarks may seem excessive as interpretations of what are, after all, only a few common words at the beginning of the poem. Is this not merely making the text a "pretext" for our own fantasies? It is important to remember the dissatisfaction with which we began. The text had set limits upon our reading, it had prohibited our access by its very construction. At that point we saw that it would not do to read the poem "simply" (*einfältig*), at a single glance. We had already trespassed against its temporality by reading it so. The poem requires the reader to read *into* it, not just *through* it. It is not an *Aussage*, the meaning does not come out suddenly all at once at the end. As a "text," it is a weaving of several strands which operate in particular temporal relation to each other, sometimes simultaneously, sometimes pro- or retrospectively. In our attempt to approach this text we must allow ourselves to be caught up in it, to be come entangled in it, and to follow the threads where they lead us.

So following, we have once again returned to the poem's beginning and have read anew as if for the first time what we had presumably already understood, but which turns out to have been completely different than we thought. We thus find ourselves, as readers, in the position of the *Landmann* himself, on this day of celebrating. For what, after all, is this poem *"Wie wenn am Feiertage...,"* if not a song of celebration, a hymn? It is important to be clear about the temporality of this movement. The *Landmann*'s "going" is not a going (away) but a returning, a coming (back); not to something foreign or distant, but to the most familiar, to *das Nächste*, indeed to himself. We might be tempted to phrase it thus: things are not what they seem. This does not mean that "reality" is somehow different from "appearance," but that seeming, outwardness, belongs to Being as an essential part of its nature, and in fact as an opening to temporality. It does not, however, exhaust reality. It merely provides a place from which to begin. Through seeming, access is attained to that which is and *in the way that it is*—namely, temporally. Thus, we say that the appearance is different from the reality, but that is only so because we have removed the appearance from temporality, frozen it in its motion, petrified it and thus killed it in order for it to become an image. What the poem here accomplishes is to create images through words which do not just say one thing, but rather contradict themselves and our attempts to turn them into static objects (concepts). The *Landmann*'s going/coming on this day of celebration which is no "day," are two examples of this process of temporalization. The word favored by Hölderlin is taken from Heraclitus: *ἓν διαφέρον ἑαυτῷ*.[19]

It has not been sufficiently noticed that the *Landmann* and the *Feiertag* both appear only once in the poem, at its beginning, and then disappear. They are seldom accorded the significance they require for a reading of the entire text. The figure of the *Landmann* returning to his field only displays its ontological character when taken together with the *Feiertag*; both actions involve a renewal of time, a new giving of time. Having provided a beginning and a point of view, they become absorbed in the world that unfolds. It is through them that the image unfolds. We will frequently encounter the figure of "returning to the same" and its identity-in-difference. Reading the text with this figure in mind will help to avoid the trap of taking its language too literally.

We have already asserted that the *Feiertag* founds or opens a certain time, which is the time of the new day in St.I, as well as the time of the poem itself. Now we must comment on the textual context in which this founding takes place. The *Landmann* goes to see a new day, a *Feiertag*, at the beginning of a strophe, at the beginning of a poem. The discourse which announces this new beginning, however, is itself no beginning. It is a comparison, as the opening words "*Wie wenn...So...*" indicate. We have to ask ourselves what purpose such a beginning *in medias res* serves. The first impression is that one has just entered a conversation which has been under way for some time. Indeed, although the factor of time is immediately made evident by this construction, it is from within time that we become aware of its existence. That is, reading the first line makes us aware that we are *not* at the beginning of a discourse, as we might have imagined, having found it at the "beginning " of a poem. We cannot, however, determine with assurance how long the discourse has been carried on, i.e., when it began. Even to presume that there is an absolute beginning is not possible. Rather, the poem quite clearly presents a non-original beginning. This does not mean that the poem lacks a structure, that one could "begin" reading it from any arbitrary point. The text itself refuses to accept its starting point as *the* beginning. The poem presents itself as though it were a part of a previous, though vanished, text. It maintains in its opening figure a living connection with that past and understands itself as a continuation and response to it. Any surmise about its thematics or point of view would be purest speculation. We are left, then, with the purely formal (i.e., temporal) fact of the (absent) presence of this prior text and the poem's connection with it. Already here we are confronted with a present textual response to an absent context.

This point should be emphasized. Although the words of the opening image (St.I–II,4) may be understood as they stand, their function only appears when we place them into a discourse which existed prior to them and from which we, as readers of this text, are excluded. The poem introduced this comparison as part of that discourse, and has used it to provide a basis (a jumping-off place) for the poem we now have. Although we may be curious about the nature of this previous conversation, I do not believe we have any justification for surmising its nature or content. Rather, its significance is to be found in its clear rejection of any absolute origin. We might describe it as a "fragmentary beginning," as a counterpart to the fragmentary ending of the text. However, whereas the latter is presumably the result of an authorial decision which could have been altered, the former "fragmentariness" is essential to the poem. It is the same kind of non-orinial origin which was found in the *Feiertag* and in the *Landmann's* "going." By refusing a claim to absolute priority, the opening figure helps to make possible the temporality of the poem.

The boundaries between past and present in this strophe are obviously quite fluid, despite the decisiveness with which it proclaims a new day. The thunder of the nocturnal storm lingers on in the present, though from a distance. Hardly has the day been announced (twice), than the poet looks back to the night which has just past. This move may strike the reader as puzzling, since the night was the time of violent eruptions of energy, hardly a pleasant time, and thus in conflict with the present, the *Feiertag*. In addition, the oxymorons *"aus heißer Nacht die kühlenden Blize"* indicate a state of affairs in which that which should be cool and peaceful is "hot," and that which incorporates violent energy is "cool." It was a chaotic time and obviously opposite to what we now think of as the "natural" order of things, in which nights are cool and lightning fiery. Yet, what I find most original in this passage is the fact that, for all of its surrealism, it remains eminently susceptible of a naturalistic interpretation. Like Dieter Henrich, I find that Hölderlin wins *"wesentliche Elemente der Bauform des Werkes"*[20] from his acute observation of natural phenomena. On summer nights, for example, the weather may indeed be "hot," and the thunderstorms which arise do seem to cool the air. This is not just an "arbitrary" conceit; it draws upon a basis of experience which is accessible to all, and yet it does not remain satisfied with pointing to that experience, but takes hold of it and transforms it into a self-contained figure of elemental forces. By the addition of the words *die ganze Zeit*, the poem expands and deepens the night into an age, possibly into the entire temporal world. Since this temporality has been set off from the present through the use of the imperfect tense and the contrast of "morning" and "night," we are left again with the impression that "now" stands outside of time, as it has been known; that it is a new beginning.

We notice as well how words begin to lose their definiteness: *wenn*, for example, appears twice in three lines and yet is different each time. Its first occurrence introduces a story, it is hypothetical. The second appearance specifies a particular time within that story. One use founds the time as fiction and one delimits a portion of the time within the fiction. Both look and sound the same. Their common appearance helps sustain a continuity in the strophe, while their divergent use (meaning) brings an awareness of diversity within that continuity. Paul Coates points to this use of oxymoron and paradox as a means of creating

> moments of fully conscious synthesis, but in so far as this synthesis is violent and momentary it leaves the rest of the poem outside the sanctuary of meaning.[21]

The next aspect of the introductory image in St.I and St.II that demands our attention is its magnitude. Many writers have commented on its size, and have pointed to Pindar[22] as a possible precedent for such a practice. Jochen Schmidt, for example, sees its rhythmic scope as an essential means for producing the hymnic style.

> Immer wieder hat Hölderlin gerade im Proemion seiner großen Hymnen ... durch gewaltig ausschreitende Satzrhythmen den Eindruck des Feierlich-Erhabenen erweckt. Es ist der angemessene Habitus der spezifisch idealistischen Dichtung, die sich auch gedanklich zum Höchsten erhebt.[23]

Only Rainer Nägele, however, has gone beyond a simple mention of its extraordinary appearance here and attempted to reach some conclusion about its effect.[24]

The practice of explaining something by relating it to a more familiar area of knowledge is ancient. Usually the protasis, which supplies the explanatory parallel, is derived from a more familiar field of knowledge than the apodosis. It may be drawn from widely-understood stories or from everyday phenomena, so that by directing attention to this more familiar area, the comparison allows the listener to make the connection to the matter at hand independently. As such, the comparison is usually brief; otherwise the conection between protasis and apodosis itself would need to be clarified or interpreted.

The poet in *"Wie wenn am Feiertage..."* does not seem concerned about such comprehensibility. As Nägele points out, he has created a comparison of such magnitude, complexity, and unity, that the reader tends to forget that it is in fact only part of a comparison, not the whole. He does not compare in order to simplify, but to synthesize, to make a complex relation between two regions of human experience. This is an interesting phenomenon, since it shows how a linguistic-argumentative process, merely by attaining a certain "size" (i.e., form), can free itself momentarily from its context and seem to exist for "its own sake." Actually, "size" is not the best descriptor here, since the important dimension is temporal duration. Human attention being temporally limited, if one approaches or exceeds that limit, the contours of discourse become blurred. This effect can be impeded by mnemonic devices which serve to create internal connections and thus unity, enabling the reader/listener to extrapolate connections also where they are not explicitly given. Such a connection is not developed in the opening strophe of this poem.

We need to ask why the poet extended the image to such a degree, what purpose did the listing of morning, night, lightning, thunder, stream, earth, rain, vine, etc., have? It could be claimed after a reading of the entire poem that these objects or phenomena are "symbolic," that they prepare for the action to come, for example, in Semele's encounter with Zeus, or in the poet's grasping *Des Vaters Stral*. As such premonitory factors, they would help "unify" the poem, acting as "preview" and fulfillment. What does the poet say about the poets? That they stand *unter günstiger Witterung*, like the trees. This is the actual point of crossing between the image in St.I and the description of the state of the poets in St.II. The favorable conditions apply to both the realm of nature (St.I) and to the poets (St.II), the former in a literal sense and the latter in a figurative sense (or is it so far from literal too, if we consider what happens to Semele?). *Günstig* is a statement about the future and an appraisal of that future from the standpoint of the present. In order to show how favorable the present is, the poet describes the *withdrawal* of the violent forces of Nature, but also the *traces* of their presence. The truly favorable aspect of the present is thus dual: the absence of violent nature and the continued presence of her effects (new growth, rain, brilliance of the sun in the rain-soaked grove).

Further, the individual elements of the comparison seem to belong to each other as natural phenomena; they do not individually transfer to the situation of the poets. It is as a total image that they become meaningful as promising good things. If we try to take them as in some way foreshadowing later events in the poem, we must be struck rather by how much they diverge. To take only the most striking examples: in St.III the poet exclaims, "*Jezt aber tagts!*" Can we identify the dawning which occurs here with the dawning described hypothetically in St.I? The latter dawn exists as part of a comparison which is assumed or projected, not as actually existing reality, as in St.III. Are the *kühlenden Blize* which fell *die ganze Zeit the same as* the single lightning bolt which hits Semele's house or the *Stral* which is ignited in the poets' soul (St.V)? They seem on the contrary to be distinct.

I do not wish to dispute the possibility of any relationship whatsoever between the appearances of these words in the poem. That they share a certain *gemeinsamen Geist* is undeniable. They are not, however, the same. They are another example of "difference-in-sameness" (*ἓν διαφέρον ἑαυτῷ*).

Before going further, we might stop and ask what the import of St.I as a comparison with poets is. It shows a natural landscape (a field) breaking out in new growth, and it shows the stormy background to this new growth. The comparison is constructed parallel to the opening verse, in that both show a strong continuity with the past. Within the landscape, the poets are to be understood as the trees which have withstood the stormy night and now bask in sunlight. Thus, they stand under "favorable conditions." In each individual example listed, the object seems to be returning to its "fundamental" form or state; thus, the stream returns to its bounds, the grass turns green again, etc. The poets, too, we suppose, are returning to their essential or fundamental state: namely, to their activity as creators of poetry. We notice, though, that they had been present throughout the storm (or else they would not have received *des Himmels erfreuendem Reegen*). The significance of the storm for the poets is as yet not clear, although it is hinted at even here.

St.II completes the comparison (indicated by *So stehn sie...*) begun in St.I, but without ever clearly stating what its subject was. The reader must read the rest of the strophe and then conclude retrospectively that it was the poets to whom the comparison referred. *So stehn sie...*, as third person plural, appears to continue to speak about the *Bäume*

Haines, who were also "standing." This impression of continuity is further supported by the immediately following *günstige Witterung*, for the trees were said to be standing *in stiller Sonne*—certainly a favorable condition. Even the "*sie*" in the following verse keeps the reader believing that he or she is still reading about the trees. One cannot know in advance, until the last word of the verse, that "*sie*" in v.1 refers to *both "Bäume des Haines" and* the poets.

The play of the ambiguity of *sie* ("she," "it," "they;" "her," "it," "them") is breathtaking, as will be seen if we try to plot out the references of St.II, 1–4.

"So stehn sie..." refers to *"Bäume des Haines"* and to *"sie"* in St.II, 2.

"Sie, die kein Meister allein..." This use seems to repeat the previous *"sie"* (St.II,1), expanding our knowledge of those standing in good weather. But it is also connected with *"die"* ("those who...") and would seem to still be the subject, but turns out to be the direct object of *"erzieht."*

"die wunderbar/ Allgegenwärtig erzieht..." This *"die"* could be the subject or the object of *"erzieht,"* since it cannot yet be determined if it refers back to *"So stehn sie..."* or forward, as "she who...educates...," i.e., Nature.

"Die mächtige...die Natur." Grammatically, each *"die"* could still be acting as a subject of *"erzieht,"* but, having reached *"Natur,"* it is clear that it is the subject of at least the last two and thus that those *dies* in 2) and 3) were direct objects.

Aside from the astonishing virtuosity with which the poet has constructed this comparison, what needs first to be clarified is the total effect of these few lines of St.II. It is already clear from the structure of the comparison that these verses are to provide the *tertium comparationis*, thus making the "point" of the comparison to which all of St.I was devoted. To do so, they need to make a clear distinction between the terms in the first part (protasis) and those in the second part (apodosis). As we have seen, though, "*sie*" (St.II,1) refers simultaneously to both sides of the comparison, thus blurring the distinction. Instead of taking the form *"like the trees after a storm, so the poets now...,"* it appears as *"like after a storm the trees stand..., So now do they who..."* In other words, the poets are not so much *compared* with the trees as *identified* with them.[25]

Secondly, the lines (St.II,1–4) which are to provide the release or satisfaction of the expectations aroused by the opening strophe instead veil the identity of the intended subject—the poets—through the ambiguous play on the words "*sie*" and "*die*." Here the poet keeps the reader guessing, shuttling between the words presently being read and previous words, future references, decisions as to gender, number, and case. Having read or heard the sentence one time through, one is still not done with it; it must be read again, now this way and now that, before "sense" can be made of it. The actual subject, however, remains as obscure as before.

This construction is far from unusual in Hölderlin's poetry. We will find further examples of it in *"Wie wenn am Feiertage..."*. Its significance, in my view, lies in its kind of temporality. It explodes the typical linearity of language into a fragmented, indecisive (at times undecidable) network of more or less interconnected linguistic units, which somehow manage to suggest a meaning. By doing so, it also brings the reader into contact with its subject, Nature. To understand this word we need to follow its use here. Actually, only at the end of the entire comparison is *Natur* heard. But that does not mean that she was not present earlier. We have already noticed the activities in St.I, in all of which Nature participates. That is, Nature has shown herself[26] in the form of storm, flood, day, growth, and sunlight. But these are particularizations, they are familiar to us through their effects. They are also temporal: i.e., they do not last indefinitely. As such finite temporal manifestations, they are similar to what the *Landmann* perceives in his everyday activity on the field. The storm is important because of the danger it represents, but also because it brings life-renewing rain. The two aspects seem contradictory but belong to one and the same Nature. As long as we see only the river, the vine, the ground, etc., we do not perceive "Nature." But, as on a "*Feiertag*," the "*Landmann*" returns to his land in order *to see it* in its essence, so now does Nature return to her bounds, to her essential state, without the distractions of particular phenomena. Like the trees, the poets do not go, they remain (*stehn*). Nature reveals herself in her glory by receding to her original bounds, after the chaotic activity of the night. Thus, she is seen as *allgegenwärtig*, *göttlichschön*, etc. We have now to look more carefully at these attributes.

Fundamentally, Nature is ever-present. She thus forms the basis of temporality, she allows the world to take place. She is also *mächtig*. We have already had some indication of her power through the description of the receding storm. From whence does her power arise? It seems that ultimately she is powerful because she encloses us (the poets); we can find neither beginning nor end to her (and yet she is temporal), and so we are in awe of her. Later in the poem we hear of her travelling in storms from the heights to the depths; such is her power.

But her power is also self-effacing. Just as it cannot be captured by individual descriptions, so too is she different from a "single solitary master," who would be tempted to exploit or persecute those in his charge. Rather, she holds the poets *in leichtem Umfangen*. The relationship between Nature and the poets is thus highly ambiguous. In spite of her power and omnipresence, she operates so subtly that one is hardly aware of her. Yet it is Nature who holds the poets; she raises them in such a way that they do not even notice her. They seem to produce their works on their own, and yet Nature is the enabling power. It is impossible to separate Nature's contribution from that of the poets themselves. Her beauty inspires the poets to perform their art. Nature is implicitly a feminine figure and as "beautiful" particularly affects the masculine poets.

On the other hand, Nature eludes our grasp. Although she is everywhere present, in her "light embrace" we do not see her. Similarly, she cannot be localized like a single "master." And, most importantly: at certain times she sleeps. Although she is "ever-present," she cannot always be reached; she is absent-in-presence. The image of "sleep" is designed to illustrate how Nature can both be all around us and yet "return" at times. Just as a sleeping mother may comfort her children, since she is manifestly still "there," yet her unresponsiveness precludes any immediate sustenance.

Sleeping is a "natural" phenomenon, familiar to all humans, in which one's mind seems to depart the body—to go away, even while one remains present. Temporality is thus essential to its understanding, often suggesting a future-oriented hope for the eventual return of the "departed" to him- or herself. The poet does not rest content with present-in-absence Nature, but goes further to distinguish different kinds of presence-in-absence:

1. *...among the gods* This in itself can refer to weather, which is one of the most changeable of phenomena, as well as to the gods as religious entities, and thus to religious faith in general. If the latter, it would be saying that religious belief at times seems to stagnate, people fall away from the old gods, are only interested in worldly things, become skeptical and rationalistic, etc. Or, it can refer to the gods themselves, who at times seem to care little for human desires or sufferings.

2. *...in the plant world* This is what we ordinarily refer to as "Nature." Here, of course, there are "times" (*Jahreszeiten*) of growth and of stasis, and yet life does not cease; it is still present, but it is present "still" (*still*), not obtrusively, and one must look carefully for it.

3. *...and among the peoples*. Here, too, one cannot say that a historical group ceases to exist or live, but at times its vital activity goes unnoticed, does not find a response in the wider arena of international affairs.

All of these relations are able to be implied by the simple figure of speech that Nature seems to "sleep." What draws us to this figure is its character of simplicity belying complexity. Nor is that all. The complexity is of a sort in which parts often contradict each other. Thus, Nature is simultaneously present-but-absent, and in fact she is so in three very different areas and thus in three very different senses of the word "asleep." The unobtrusive word "*unter*" connects the various types of differentiation and also provides a further example of simplicity belying difference. It is another example of the kind of polyvalence shown by "*wenn*" in the first two lines. In the last segment of the metaphor of sleeping Nature we see a reference to history: at various times a historical people may seem to be dormant, inactive, unchanging. Here, however, it is not the peoples who sleep, but Nature who sleeps under/among them. Clearly, individuals continue to be born and die—i.e., change continues—but, since the society as a whole in its institutions remains the same, nothing *significant* changes (*plus ça change...*), and thus time seems to stand still, as when one is asleep. We must ask ourselves what this can mean: *"Nature sleeps among the peoples."* Since Nature is the personification of the growing principle of all things, if she sleeps there can be no growth.

Nature is thus complex and highly ambivalent. She is both powerful and yet holds the poets "lightly." She is "everpresent" and yet "sleeps" at times. She forms the basis of temporality through her changeability and multifarious forms, yet she is "older than the ages." She is far broader as a concept than ordinary usage would imply. As Richard Unger has shrewdly pointed out,

> It is remarkable that no image is evoked in this strophe's attempt to name Nature; there is no figurative language because, clearly, if the poet's word is to be "the Holy," it must be unevasive. It should not seek to interpose the figure of some particular entity between itself and universal Being.[27]

I would venture to suggest that Nature's essential quality, that she is *"wunderbar/ Allgegenwärtig,"* is represented more faithfully through the poetic syntax of St.I and St.II than through any "image" pictorially understood. Specifically, she is the basis of the gods, of the vegetative world, and of human history. How can we thus answer the question, "Does Nature influence the poets or not?" Nature does not influence or educate the poets like a "single master." A particular individual educates by teaching particular facts or methods. Nature educates in a non-particular, non-materialistic way. She operates by allowing things to come about. When the poet experiences the coming-about of things, he learns not just about them, but through them about Nature herself.

It is also significant that, immediately following the panoramic and promising (*günstig*) description of the situation of the poets in Nature, the poet shifts to deal with the "negative" or opposing sides of what he has just proclaimed. Throughout the analogy of the St.I–II, the poet described the *present* situation of the poets under Nature's tutelage and declared it to be good. He now abruptly (in the middle of St.II) turns back and offers an explanation or exculpation of the poets' *trauern* through the device of Nature's "sleeping." The poets are now standing in good conditions, *"Therefore...they appear sad..."* In St.I,3–4 we read:

> Ein Landmann *geht*, des *Morgens*, wenn
> Aus heißer Nacht...*fielen*
> Die ganze *Zeit*... [emphasis added]

In both cases the poet moves from the present, in which things are looking hopeful, promising, and thus future-oriented, into the past, to the violence of the night or the sadness of the poets. This return interrupts the linear flow of the poem and keeps the future distant. St.I shows Nature coming back to life; St.II describes the favorable state of the poets within Nature. It would seem natural that the poet would then proceed to describe what will emerge from this situation, but that does not happen. Not that which is coming, but that which is here *now* needs to be seen in its historical depth. Only through the (absent) past can the present be understood as what it is. All-present Nature is also present-in-absence. In fact, she must go away, in order that people can become aware of her all-pervading presence. Even when she is "resting," she "intuits" (*ahnt*). These words describe ongoing processes, not single actions.

The point of interest here, that Nature only "seems" to sleep, seems intended (for the reader) to maintain an external perspective on these events. We notice 1) that Nature "seems" (*scheint*) to sleep; we do not hear that she does in fact do so; and 2) she appears (*scheint*) in different places and times—the action of sleeping is removed from the specific agents (gods, plants, men) and placed under the heading of "Nature." This is shown first as Nature "appearing" to sleep, and second, as the poets' "face" showing sadness (*trauert*). Both words refer to external appearance[28] (as opposed to "*leiden*" or "*schmerzen*," for example). We already know the poets are Nature's pupils, so it is reasonable that they should be sad when their preceptress is absent; however, we are told they only *seem* sad, she only *seems* to be away, they only *seem* to be alone, and that they are actually still in touch with her through *ahnen*. This last word deserves a few words as well. Through *ahnen* one is open to the future; in fact, one has contact with the future, but it is a hidden contact, not capable of being truly shared with others or justified before them. The point here is that 1) the poets are not alone, because they are in (paranormal) contact with Nature; i.e., they "know" she will return and they await her; and 2) the future, which is usually conceived of as something *different* (both from the present and from the past), turns out to be a *return of the same* (past), and 3) Nature does not truly rest, because her resting is also an "*ahnen*."

Already in the next line the *ahnen* is fulfilled: "*Jezt aber tagts!*" The return to the present presumably puts an end to the *ruhen* and *ahnen*. And yet not quite, for the poet immediately turns back to the past ("*ich harrt und sah es kommen...*"). The present here, just as in St.I and II, makes a brief appearance, but then recedes before the past. There is, however, a certain progression: in St.I the present was hypothetical, clearly distinct from the poetic present; in St.II, however, the NOW (*"So stehn sie..."*) was presumably the poetic present but with a strong future orientation, which carried through the strophe into the final *ahnen* and *ruhen*; in St.III the present is emphatically "here," but only for a short while.

A similar progression can be noted with reference to Nature: in St.I she was present, though unnamed, in particular manifestations; in St.II she was named and praised as the protectress and educatrix of the poets, but then withdrew into sleep, leaving the poets isolated. In St.III she returns once again, accompanied by the day, and now in even greater glory. She is not just *"göttlichschön,"* but *"über die Götter des Abends und Orients,"* and indeed seems to be responsible for, if not identical with *"das Heilige,"* the essence of the Holy. But to see that, much time must be spent on these two rather inconspicuous lines. Their apparent simplicity hides an astonishing complexity.

1. *ich harrt...* What do we understand by *harren*? It has to do with waiting, i.e., with remaining in one's place, not foresaking the world to which one has been given, and with a certain degree of resistance (notice the homonym *"hart"*) to be overcome. It carries the implication of withstanding the temptation to give up, give way, wander, forget. It also implies that one intentionally neglects one's own concerns for the sake of remaining here. The poet thus achieves a certain resistance to temporality, by not passing away but instead remaining. The action involved is both "active" and passive; the poet must exert himself to stay the same.
   Some things can only be perceived by such *harren*. They require the passage of time as played against a (relatively) stable background. Such a thing is "dawning," the approach of day. It occurs slowly, progressively, seemingly "behind our backs," when we are not looking. So too does a new historical era come into being: at first, only dim outlines of new forces can be made out, but they grow progressively more distinct—if one waits. Thus, patience or persistence is an absolute necessity, if such temporal phenomena are to be perceived at all.

2. *...und sah es kommen...* Actually, this line represents two actions: "it was coming," and "I saw it (as it came)." It also carries the figurative meaning, "I foresaw it." That is, through *harren* and *ahnen*, I gained a kind of knowledge of the future. The syntax of the German language requires the poet to say *"es,"* with the result that he saw a "thing" doing something. In fact, what the poet is speaking of is a process, and he de-

scribes it as an "approaching," which we would ordinarily call "dawning"—a mystification, insofar as it merely pushes the identification of what it means to "dawn" one step further back. The poet is describing the observation over a long period of time, in the past, of an approach—of what? What is *"es"*? It could, of course, simply be refering proactively to *das Heilige*, but this seems to cover over too quickly the profound ambiguity contained in that term (see below and *"sie"* in St.II,1–4.)

3. *...und was ich sah...* This is a paraphrase of the previous statement that the poet did indeed see something. Its significance comes from its rhythmic similarity to 2) and the repetition of *sah*. It helps sustain the backward-looking anticipation.

4. *...das Heilige sei mein Wort.* We notice that there is no formal syntactical connection between 3) and 4). Taken simply, literally, the poet expresses the desire for his word (his poem) to be "holy" or "the Holy." But we cannot take his word so simply, because it obviously refers back to 3), which itself is no "thing," but a process of coming.

It seems the poet is expressing a somewhat ritualized greeting or naming of what has been revealed to him as "holy," or at least what he has decided to call "the Holy." Much ink has been spilled over the question of whether he does in fact (or only in text?) *name* the Holy,[29] but one thing is clear: he does not "name" in such a way that the reader can with certainty say "what" it is or that it is a single entity. The poet seems to be referring to the coming of Nature and the day, especially in view of the next lines. But their coming only indicates a reason for his exclamation; they do not clearly identify what is "the Holy." That "Nature" is meant could be surmised from what the poet has already said about her, yet she is so far from a single thing as to explode the very meaning of "entity." Further, her elusiveness finds appropriate expression in these lines, which seem to refer to a "thing" (*es*) which finally does not appear. Indeed, if we take the lines as referring back to the dawning which immediately precedes them, they then celebrate the emergence of the "world" (cf. the *Landmann*) as for the first time.

As a whole, then, the poet sees the coming of day, which is the return of Nature to activity, and can find no better word to describe it than "holy." This is not to say that the poet thereby "names" or "founds" the Holy. If anything, he destabilizes the Holy by pointing away from any particular ineffable something. In fact, the structure of the statement shows that his response is not to any identifiable "thing" (not even, strictly speaking, to Nature), but to the sudden awareness (*"tagts!"*) of a process that has been going on for some time.

In a move that occurs frequently in Hölderlin's poetry, the poet has distanced his words from his subjective feelings. He does not say, *"What I saw was holy,"* or *"I saw the Holy,"* or even *"As far as I'm concerned..."* He replaces himself and any personal impressions with his "word." It may be a single word of greeting or it may include his entire poem. In a similar fashion, the "Holy" is a substantivization of an adjective; it is that which we describe as "holy," that which evokes the exclamation of "holy." Clearly, this definition is circular and testifies thereby to the "emptiness" of the concept—which does not dispute its existence and effectiveness (*Wirklichkeit und Wirksamkeit*). In order for us to understand what has caused this exclamation, the poet describes the magnificence and scope of Nature.

The rest of St.III restates the fact that Nature and Enthusiasm have awakened. The poet brings out more explicitly what was indicated in St.II: Nature is not only *göttlich*, she is superior to the gods, the individualized representations of the Holy. She encompasses all times, as the precondition for the emergence of temporality. The poet emphasizes through repetition (*"Denn* s i e, s i e *selbst..."*) that Nature has indeed awakened; but the duplication splits, rather than reinforces, the identity he seems to be asserting. In fact, it first *creates* the split between "authentic" Nature (*sie selbst*) and some other. I believe the emphasis on *"selbst"* is to be seen in connection with her being *"älter"* and *"über"*—stressing the transcendental Being of Nature, which cannot be captured in any single being, not even in a god. *Sie selbst* seems to contrast with a partialization or fragmentation of Nature, similar to the *Meister* mentioned above. It is not just a part of Nature that has awakened, it is Nature in her totality, in her essence, which is temporal, who has returned to activity.

The structure of this strophe (especially verses 3–9) again keeps the reader from being able to identify who is being discussed until it has been completely read through.[30] The initial *sie* is left unspecified until, as a kind of third repetition, *Die Natur* is spoken of in St.III,5, in the middle of the strophe and of the poet's description of her. Having named her, though, the poet goes on (*"Und hoch vom Aether..."*), seeming to expand his eulogy of Nature, whereas it turns out in the next-to-last line that he has been speaking of *die Begeisterung*. Thus syntax, instead of serving to unify and solidify the text, is a means to suggest more than one thing at the same time, to combine contradictory things into a complex whole. The same device was discussed above, regarding *sie* in St.II.

"Die Natur ist jezt mit Waffenklang erwacht." This is one of the places in the poem which seem to refer to the military events of the poet's (Hölderlin's) day.[31] Although *Waffenklang*, if taken literally, would indeed seem to refer to battle, what evidence is there within the text that this is indeed the message? Most of the commentators simply take the equivalence of *Waffenklang* with revolutionary actions for granted. Richard Unger, for instance, refers only to *"the cataclysmic event of the French Revolution and the ensuing wars which many of Hölderlin's generation considered portents of apocalypse."*[32] Even if Hölderlin himself considered them thus, the question remains open: does this passage about Nature refer to them? Peter Szondi,[33] however, cites two other poems of Hölderlin to show that he did expect a transformation of man's relationship to the divine out of the upheavals of the later eighteenth century. Yet it is significant that he must turn to *other texts* to make this point. It would be reassuring to have more specific support within the poem for such a view. The poet, after all, is speaking of Nature, who is *"Allgegenwärtig," "göttlichschön,"* etc.—not of *"Völker,"* for example.

Further, is Nature awakened *by* the sound of weapons (i.e., at the same time as weapons sound, *"mit* der Zeit *des Waffenklanges"*[34]), because nations are once again at war, or is there *Waffenklang because* she is now awake? That is, does she herself approach with the sound of weapons, as in storms? The first possibility seems senseless, for how could mere men, no matter how many, awaken something as powerful and elusive as Nature, who follows her own rhythms? Besides, we have already seen the poets helplessly waiting when she sleeps; if they could have done something to awaken her, why did we not hear of it? The second possibility seems to refer to the tremendous impact of Nature's awakening on the poet, similar to the coming of day in St.I. Seifert here goes even further by insisting that

> Die Natur erwacht nicht unter begleitendem Waffenklang oder wird durch diesen erweckt; sondern sie kommt im Waffenklang oder als Waffenklang

zu sich und in die Offenbarkeit.[35]

[italics in original]

I find little essential difference between this interpretation and the second, above. Both emphasize Nature's power and impressiveness.

Regardless of the interpretation of Nature's relation to the sound of weapons, one thing remains puzzling: how is this description of her awakening accompanied by the sound of weapons reconciled with the image we were given in St.I of daybreak (i.e., of Nature's return) as a return to peace and order after a violent night of storms, lightning, floods, etc.? Previously, the *günstige Witterung* referred to the trees standing *glänzend/In stiller Sonne*, whereas now Nature and Enthusiasm dynamically fill the entire universe, *"hoch vom Aether bis zum Abgrund nieder."* The description of Nature in this strophe complicates the picture. Contradictions belong to that picture, for she can be both mighty and quiet: she can approach so gradually that it escapes our notice, until all of a sudden her presence startles us like the "clang" of weapons.

This seems an appropriate place to call attention to the undertone of violence and aggression that accompanies the *Waffenklang*. In St.I we learned of *Blize*, which were actually a relief in hot nights and anyway were safely past. We also heard the thunder which still sounded, although from a distance. Thus, violence was not completely lacking in the opening image, but it was relegated to a safe distance from the "present." In St.II the poets were said to be in a favorable situation, partly because they owed allegiance to *"kein Meister allein."* The threat of a superior and guiding force was excluded from reality, merely existing as a possibility which (fortunately) was ruled out *for the present*. Now, however, in St.III, Nature herself is accompanied by *Waffenklang*. At this point it is not important to decide whether she is responsible for the weapons and the battle or strife that they imply. What is more important is the heightened visibility of danger and its connection with the source of protection, Nature.

The language of this passage is studded with paradoxical juxtapositions; as for example,

(III,3) *älter denn die Zeiten*

(III,4) *über die Götter des Abends und Orients* —Nature transcends all gods, whether occidental or oriental. Notice also the contrast between *über* in the sense of vertical (in space as well as in hierarchy) displacement, and in the sense of being "above" the gods, who are usually above all beings. The accusative here (*über die Götter...*) is also somewhat perplexing: is Nature "placed" over the gods (accusative of motion)?

(III,5) *Natur...Waffenklang* Nature is accompanied by human technology; that which is peaceful and nurturing together with the warlike.

(III,6) *hoch vom Aether bis zum Abgrund nieder.* The poet has here brought closer together and actually combined the opposites of height and depth by placing them in the same verse.

(III,7) *Nach vestem Geseze...aus heiligem Chaos.* Even in Chaos there is fixed law. The connection between *Chaos* and *Abgrund* stems from the Greeks, who seemed to have considered chaos as a *"weite Kluft, Schlund, Abgrund."* [36] (See below, Chapter Six)

(III,8) *fühlt neu...sich...wieder.* Conflation of newness and repetition.

One of the above items deserves more extensive comment: *aus heiligem Chaos*. In the same strophe the poet applied the word *das Heilige* to what he had seen; here he connects *heilig* with "Chaos." The latter is the Greek term for *"that which came into being before all else,"* according to Hesiod.[37] In Ovid, the second *locus classicus*, it is a *"rough mass,"* in which everything is without order.[38] It seems that when Nature returns to herself ("awakens"), she emerges from Chaos, which is "holy" by virtue of having preserved Nature ("whole") during her absence. It remains paradoxical, however, in that Nature's return to activity is accompanied by the clash of conflicting elements, and thus has a portion of "Chaos" herself; whereas her emergence from Chaos is *nach vestem Geseze* and thus predetermined, i.e., non-chaotic, non-random. As many writers have observed, *Chaos* for Hölderlin is not the biblical first step of creation,[39] but the *schöpferischer Urgrund* of life.[40] In fact, Wolfgang Binder sees Chaos, for Hölderlin, bringing forth time itself.[41] All that still leaves us with Nature as the origin of all things—but she herself comes from Chaos as preserver of the origin of all things. It seems this contradiction also occupied the Greeks themselves.

The approach of Nature, signalled in St.I–II, has been accomplished in St.III. She has arrived in all her majesty. The time has come NOW. This means there is no further waiting, but also no possibility of escape. The poet has seen the approach of day and has called it "holy." "Holy" is a word usually reserved for a god, signifying that which is proper to the god and thus removed from human influence. It is foreign (*fremd*) to humans. In this case what the poet has seen, the Holy, is a process of coming to light (*"tagts!"*), of coming into view. When day first begins to appear it is almost imperceptible, more of a feeling (*Ahnung*) than something sensory. When day has arrived, it is not simply the presence of the sun in the sky, but rather it brings all things into view. It gives everything its form and color, equally, without distinction. This overwhelming, overflowing, into-presence-bringing, is the approach of Nature, which is actually a return. (Cf. St.I, *"ein Landmann geht..."*) She has not really been away, though men were not able to reach her. Her approach has been iconically represented in the poem through the hesitancy of mentioning her name, through burying her in complex syntax and referring to her in the third person, until finally the poet burst forth with her name, and even added that "she herself" is here. Her approach has been as inexorable—i.e., uninfluenceable by humans—as the coming of day. She is described principally in positive terms, but should she ever turn against humans, they would be utterly in her power or at her mercy.

If we take the return of Nature as the cause of the joyous, expectant mood of the poet—corresponding to the poets' sadness at her "sleep"—and we see the approach of Nature as a setting-in-movement that proceeds throughout the whole universe, then we must ask what distinguishes this activity from the activities of the night that has passed. There were thunderstorms, rain, floods, and the grass withered; but otherwise one could say with equal justice that the night before the *Feiertag* was also characterized by the activity of "natural" phenomena. And, indeed, the poet began by showing how the results of the night's activity are actually good—helpful to the ground and the grapevine. The lightning cooled the hot night, the stream quenched the ground's dryness.

By claiming that now Nature has awakened and now we see activity throughout the universe, the poet awakens expectations in the reader that previously things were different, i.e., that things were "asleep," motionless, in isolation, etc. This does not seem to have been the case. What has actually happened is that the poet has become aware of the *temporality* of Nature. She is more than even gods and can seem asleep at times, but only because one does not have the patience to wait for her. The poet, however, has waited and has seen. And what he saw took his speech away from him and took it to itself, so that only NOW, as it begins to dawn, can he find a word for it. That word is *das Heilige.*

The Second Triad (St.IV–VI)

The second triad picks up the discussion of Nature and transforms it at the same time. Nature is still there, but only as the ever-present background; she is not named, but influences the poets and their creation, the song. That is, Nature works now through men and through the poets' art. *"Wie wenn am Feiertage..."* states clearly that the poet is animated by *Feuer*, that the *Kräfte der Götter* come forth in song, if it has grown from fiery Nature (*der Sonne des Tags und warmer Erd*). The former external perspective is now transformed and makes visible the "interior" or *geistig* sphere. The movement inwards is clearly represented by the comparison with the past, when the fire was in the man's eye (i.e., visible to the outside world), and the present, when "it" is to be found in the poets' soul. We recall how the external perspective was preserved in St.II, where the poet insisted on *scheinen*, *Angesicht*, and *trauern*.

What did such externality accomplish? It directed the reader's attention to the distance separating him or her from the poem, or from Nature. The poet described only the external aspect of the seasons and of the poets. These became "phenomena," to be observed from a distance. They were separated from us by an "abyss." As long as the abyss separates us from Nature, we can only follow Nature's appearances and cannot really understand her; but that is acceptable, since she still captures our attention, she still leads us to stay and rest and wait for her return.

I understand the "fire in the eye" as a consumption by enthusiasm which can be seen (read) in the eye. Curiously, the eye does not see here, but shows. Do "fiery" eyes see better than ordinary ones? Perhaps not, but they do give evidence (*zeugen*) of internal possession, which urges one on to greater deeds: *"wenn hohes er entwarf"* implies "great plans," i.e., human inventions or discoveries like agriculture or great cities and temples (*hohes*), as well as feats of daring. It also suggests the danger of *hybris*, in which "casting above" one's station is interpreted as an attack on what rightfully belongs to the gods. Danger also lies hidden in "fire," although not commented on explicitly. The motif of dangerous fire stands out here (although it seems incidental to the central meaning of the passage, just as the violence in *Waffenklang* was easily overlooked) because it continues an ominous subtext which will come into the spotlight (*Brennpunkt*) at the poem's end.

The man in the phrase *"Wenn hohes er entwarf"* is also related to the *Dichter* in the sense of *poietes*[42] (i.e., "doer," "maker," *homo faber*). Previously (St.II) he was shown to be dependent on Nature's presence for his own activity. Now he is seen to derive inspiration from *den Thaten der Welt*, apparently a reference to historical events.[43] In both cases, although the poet is the active, productive agent, he is seen as relying on Nature for the original possibility of producing anything. What St.IV,1–4 seems to be saying is that the poet today takes his enthusiasm directly from events in the world, just as he used to be inspired whenever he made any truly great plan or discovery. The present thus shares with the past an ability to allow man space for creation, but differs from the past in locating the proximal source of that inspiration in the realm of human events (*Taten der Welt*).

The comparison with the past continues through the entire fourth strophe. Before, when one only saw the external deed, the true nature of things was unknown; now that the fire has been lit within the poets' soul, the work of the gods is recognized for what it is: their friendly inclination to support human beings. "Fire" does not simply bring into motion; it also sheds light on what had been in darkness.

What is particularly important in these lines is the continuance of the past into the present. The gods were active "before," but were not recognized; now they stand "revealed." They have been there all along, helping man without his knowing it, similarly to Nature herself, who is present even when she "sleeps." Of course, the parallel that presents itself here most forcefully is with St.I. The *Landmann* who went to "see" his field in the morning is set next to the *Mann* with fire in his eye, the *Feld* and *Boden* reappear as *Aker bauet*, etc. But most importantly, St.I also presented a moment of revelation after a period of darkness and ignorance and showed Nature returning to bounds, to a more favorable relationship to men, just as in St.IV the *Kräfte der*

Götter are seen as *Allebendigen*, as *lächelnd*, and as *bauen*, in *Knechtesgestalt*. Again the accent comes on *erst jezt* (cf. St.III,1): NOW is the time of revelation, when we can see what was "hardly felt" before. "Seeing" here has the meaning of seeing with the mind's eye, of understanding, as opposed to simply being conscious of a feeling (sentiment). It has always been considered the preëminent human capacity. Thus, only NOW do men achieve the full measure of what they can be. With this new, more interiorized faculty of sight, man can now see that the *Aker* here is different from the *Landmann*'s "field:" whereas the field became a symbol for the man's whole world and identity, the *Aker* is to be understood as the realm of "agriculture," as general human activity for his self-support, though not necessarily as the basis of his identity, since the *Landmann* is not spoken of. In fact, the subject of the recognition is not clearly identified.

We must also call attention to the circumstance that, although the forces were in the *form* of *Knechte*, they were always *in fact* gods. This is a little puzzling—how are we to understand *Knechtesgestalt* if applied to either agriculture or general history? Does it refer to the nutritive agency of food which humans did not perceive as coming from the gods? In any case, it seems chronologically reversed: from the earliest times, humans have attributed their life and nourishment to the gods (whence the almost universal practice of sacrificing to ensure a good harvest or hunt). Only in relatively modern times did men start replacing divine agency with "naturalistic" (i.e., scientific) explanation.

St.V begins with a question: *"Erfrägst du sie?"* This question implies a doubt about what has just been proclaimed as the "revelation" of the "divine powers" in human history. It is presumably directed at the reader, though it could also be a self-questioning of the poet after such grandiose claims in the previous strophe. Further, despite the "revelation" of the gods for what they truly are, the poet here points to the song, in which their *Geist* is to be found. Just as we earlier saw the play of hidden versus exposed, so too here it is not immediate presence that the gods permit; rather, mediated through song, one catches a glimpse of the *Geist*. In this way they are separated from us by two steps: first, through song, then through their *Geist*. It would seem then that their revelation is somewhat less than complete.

Previously, the gods were anonymously responsible for human support and nourishment. *Aker gebauet* and *Knechtesgestalt* point clearly to activity in the realm of "nature" (fields, crops, etc.). Now their activity has been carried over into song, just as the "fire" was carried inside from the man's eyes into the poets' "soul." Song itself, however, has been "naturalized:" only if it proceeds from Nature (*"der Sonne des Tags und warmer Erd / Entwächst"*), does it contain the spirit of the powers of the gods. The poets, producers of songs, are still dependent on *"Die mächtige, die göttlichschöne Natur,"* and she alone ensures their divinity.

The sun brings the day, the *Feiertag*, on which the poets come forth with their song. It is thus time-founding and poetry-founding. The sun has also given itself to the earth, thereby warming and enlivening it. From this fruitful union the songs issue, but they also require the storms. Storms also come from the sky, but they do not bring warmth; instead, they bring rain. It should be noted that in most circumstances *Wettern* are *opposed* to the sun: they block the sun's power and even its visibility. *"Wettern"* takes on particular significance also from its position in the verse; separated from the former two components, *"Wettern"* themselves are split into the physical kind (*"...die in der Luft"*) and "others." The chain of causality does not want to be broken until it has returned to its origin (in heaven and earth). Storms are transitory things: they come unpredictably, not like the sun or *Feiertage*, and they can be violent, making them suitable vehicles for metaphors of war. The metaphor of war as storm carries over the production of massive, violent change through the sudden appearance of innumerable small "drops" (components). Both are totalizing phenomena, unpredictable in their specific manifestations but usually having the same result each time. The "other" storms are left unspecified, like the coming of the recesses of time (*"in Tiefen der Zeit"*), which does not prevent them from being "more prepared, full of meaning, and comprehensible." *"Vorbereiteter"* indicates their adherence to a plan, an order, a *Schicksal*; *"deutungsvoller"* is a pointing towards something, but also indicates a subordination to a plan, although it does not say how understandable to men that plan may be; *"vernehmlicher"* fills the gap by stating that we can "perceive" what is said. (Nowhere does the poet say that we can finally *understand* what is said.) *"Tiefen der Zeit"* are understood as distant reaches of time (cf. *die ganze Zeit*), also as history, and as connected with "profundity," (*tief ... unergründlich*). *"Tief"* also contrasts with *Luft*, as parallel to *Aether* and *Abgrund*.

We notice here also the surprising shift in meaning of *"unter:"* *"zwischen Himmel und Erd und* unter *den Völkern."* The poet had performed a similar shift in St.II,6, (*"Am Himmel oder* unter *den Pflanzen oder den Völkern"*) from "*unter*" meaning "under," "below," to "*unter*" meaning "among." In itself it does not appear significant, but it is one more element in a continuing process of undermining the supposed stability and univocality of language (cf. *"Wie wenn..."*). The sentence began as a question, which was immediately answered; the subject of the answer (*Lied*) was next itself qualified (*"...wenn es ...entwächst"*), seemingly in twofold but actually in threefold fashion (*Sonne/Erd/Luft*); the last term of the qualification (*Wettern*) then itself being qualified (*...und andern*), followed by a relative clause, again with a verb not placed in final position, as would be expected, but between the three complex, comparative adverbs and three prepositional phrases, two depending on *zwischen*, one on *unter* (which is equivalent to *zwischen* [among]).

Adding to the complexity of this construction is the temporal situation it describes. It begins with a hypothetical question in the present, followed by the statement that the song *entwächst*—this could mean that the song has "grown" out of such elements and is to be regarded as finished, as "fruit" (cf. *Weinstok, Pflanzen, Aker*, and below *Frucht in Liebe geboren, Frucht des Gewitters*) or that it continues to be rooted in and grow out of these elements (like the *Bäume des Haines*). The focus of the statement (and of the strophe) changes from *sie* (*die Kräfte der Götter*) to *Lied* (briefly), to *Wettern* (first as meteorological, then as cosmic phenomena). That is to say, the song, bearing the spirit of the divine powers, grows/has grown in certain circumstances: it is also nourished by storms which arise out of the "depths of time," i.e., long ago, and which have an intermediary function between heaven and earth (mortals). Although they have been carefully prepared and are full of significance, they are sent on their way rather casually, as *Hinwandeln*—their specific times of arrival or departure being left vague. The process of travelling, indicated by the present tense, is seen as continuous.

After this long and involved explanation for the presumed difficulty in recognizing the "divine powers," St.V "closes" (but really opens onto St.VI) by re-formulating its answer. In place of the single, poetic formulation, we are shown where poetry ends and begins (in *"der Seele des Dichters"*). This figure (*Des* G*emeinsamen* G*eistes* G*edanken*) shows by the repeated /g(e)/ phoneme its unity-in-difference (*Gemeinsamkeit*). It also serves to unify all three strophes of the second triad. Through the idea of *Geist* as the inner essence of the divine powers, the poet's soul is inflamed and inspired (*Geist* = *pneuma*); that they come to end in the poet's soul connects with the earlier statement in St.V (*"im Liede wehet ihr Geist"*), as well as with St.VI, where *sie* (the poet's soul) twice gives birth to "fruit." It should be noted here that *endend* is a present participle, indicating a continuing state of activity; a contradiction between form and content (a never-ending "end"). *Des Gemeinsamen Geistes Gedanken* are *"still" endend*; the massive, thunderous *Wetter* have dwindled into thoughts, which in turn have almost come to rest. Similarly, when Nature awoke (St.III), the phenomena ranged from the *Waffenklang* of the universe, from highest *Aether* to the *Abgrund*, to the interiorized feelings of *Begeisterung*. The action portrayed at the very beginning of the poem also suggests such a progression from the violent storm of *die ganze Zeit* to the distant thundering, ending with the trees standing*in stiller Sonne*.

It has already been remarked[44] that the contrast between the end of St.V (which is no end) and the beginning of St.VI (which is no beginning; cf. St.I) is quite pronounced (*still....schnellbetroffen*). The transition is not only abrupt, but it implies a causal connection (*"Daß..."*). This should reinforce the notion advanced above that the *"Gemeinsamen Geistes Gedanken"* find their "end" as *telos*, not just as their cessation, in the poet's soul. Although they are described as a process, and thus as gradual or continuous, the soul is struck all of a sudden. This is perhaps a kind of "ripening" which ultimately leads to fruit falling from the tree. We do not learn immediately what has struck the soul, nor what the consequences are. In fact, the soul has been struck and the cause is linked to "*still endend*," before we find out what, who, or why. In addition, "*sie*," which is grammatically the "subject" (doer) of the sentence, is in fact an "object" (done-to) in more than one way:

1. She is the effect of a previous cause (*"Daß..."*).

2. She is *schnellbetroffen*. Notice that the word conveys iconically the action it describes (speed and force); it combines with *"sie"* into one unit ("she-insofar-as-suddenly-struck").

3. She is *Unendlichem / Bekannt....* Note the reversal of the normal relationship: instead of a person "knowing" an infinite being, she is "well-known" to him/it.[45] The paradox continues with *"seit langem,"* since an infinite being would have known her for an infinitely long time. This is, of course, a fore-shadowing of the relationship of Zeus and Semele.

4. She is made to tremble by *Erinnerung*. Trembling is ordinarily intransitive, but the addition of the qualifier (*"von Erinnerung"*) brings out its passive-receptive aspect. Although the soul is well-known to the infinite being, she must now herself remember (or remember herself?), and this leads to "trembling, quaking," as in an earthquake. As J. Schmidt points out, this *Erinnerung* corresponds to the Platonic *anamnesis*.[46] In Plato, of course, the soul actively seeks to recover its memory of the *eide* (cf. *Meno*, 81B, for example).

5. We are not at first clear who the referent is. It could be the *Gemeinsamen Geistes Gedanken*, but turns out to be the "soul." The statement continues and brings her "objectness" more explicitly into the foreground, by replacing *"sie"* with *"ihr."* This, however, merely compounds the uncertainty that was already felt about her identity in the first place, now at a second remove.

6. *"entzündet"* (Cf. St.IV,4: there fire in the poets' souls was also connected with memory [*was zuvor geschah*]). *Kräfte der Götter* are also fiery.

7. *"Die Frucht...geboren..."* Notice how gender changes in apposition. *Der Gesang* is on its way to becoming *Bacchus*. Semele recedes here in favor of that which she will bear.

8. *"...in Liebe..."* The question is: whose love? The soul's for God or for *des Gemeinsamen Geistes Gedanken*? Or vice versa (God's for the soul)? Or both?

9. *"damit er beiden zeuge..."* Strictly speaking, this refers to the *"Götter und Menschen Werk,"* but that merely obscures the fact that it is the *Seele des Dichters* whose "labor" brought forth the fruit. Also, *"zeugen"* is used here in a dual role: a) as "bearing witness," and b) as "procreating" (*erzeugen*). Although this latter sense is "nonsensical" on the face of it, since the song is produced by men and gods, rather than itself producing them; yet, on further reflection, it appears to have something to it. For, one could say with equal justice that song first creates the very identities "men" and "gods." Thus,

"the child is the father to the man." Cf. *gezeugt* in St.III,7. Who was procreated, Nature or Enthusiasm?

10. *"glükt."* the "song" is apparently the subject here, but the referential object is still at least partially *"ihr."* The double meaning of *"glükt"* is also important: the song both "succeeds" and "makes [whom?] happy."

After listing all of the ways in which the grammatical subject is turned into an object, we must now turn to consider what this accomplishes. In ordinary language, when we say that a poet creates a song, the poet is the independent, self-sufficient source which "produces" the object, the song. This view is radically subverted in *"Wie wenn am Feiertage..."*. Not only is the poet no longer a unified subject, since he is split into "soul" and whatever else; his soul, too, "does" nothing, but rather is "done to," first by *des Gemeinsamen Geistes Gedanken* and then as passive subject "from" whom song is born. Jochen Schmidt has emphasized that Hölderlin's concept of poetry did not take all of its inspiration from the *Geniegedanken* of the *Sturm-und-Drang* period, but also carried over the passive nature of *Empfindsamkeit.*[47] Clearly, this passage does not show the poet heroically *"forming men according to my image,"* as Goethe described it;[48] rather, the poet is here distinguished by his heightened *receptivity* to the divine. Thus, the notions of self-identity and causality are thoroughly challenged. This development will become crucially important towards the end of the poem, where the poet's "guilt" will stem from his *eigen*händ*ige* Hand*lung*.

Together with this clearly passive relationship between the poets (or their "soul") and either Nature, who raises and protects them as well as providing the right time for their appearance (in the sense of *Erscheinung* and *Angesicht* [St.II,7]), or with the gods, who seem to take Nature's place when it comes to producing their song, this strophe also introduces the idea of "love" (*"die Frucht in Liebe geboren"*). The gods had shown themselves kindly disposed towards men (*"die uns lächelnd den Aker gebauet"*), but everything had proceeded according to an impersonal *vestem Geseze*; i.e., the forces of Nature and Spirit, which seemed so powerful to humans that they could accomplish anything. But of course one thing they could not "accomplish" or "produce" was love, and it is ultimately through love that the song (*Frucht*) is born, gives witness, and succeeds (*glükt*). At this point the imagery shifts subtly to more anthropomorphic and less naturalistic expression.

In many ways this last proclamation of the birth of song in love marks the emotional highpoint of the whole poem (even more than St.III,1). Up to now the "mood" of the poem has been almost uniformly hopeful, positive, even joyful. The first triad sets out the return of Nature and her wondrous qualities or powers. The second triad has shown how the collaboration of Nature, men, and gods has always existed but until now without human awareness, and how song, the culmination of such collaboration, comes forth at last.

The change which begins to take place immediately after this point (St.VI,5) is therefore all the more striking. Having accomplished the primary task of the hymn, namely the *Feiern*, the poet now begins to introduce defensive considerations into the idealistic schema he has laid out for us. Of course, this is not the first time dissonant voices have been heard in the poem. The fluidity of time-values and identity, the reversal or blurring of cause-and-effect have worked to undermine on a structural level the content of what has been said about poets, Nature, and time.

"So ... fiel," coming directly after the "successful birth" of song, seems intended to introduce a comparison, like the *"Wie wenn..."* of the poem's first verse. The poet has just described the genesis of song; a comparison taken from literature would seek to exemplify the process, perhaps using mythological or symbolic figures. Such a device is also familiar from the Bible, where parables are extremely common. We must, however, be cautious when reading such parallelisms in Hölderlin; they are often "imperfect," or, to put it differently, they often introduce new elements under the guise of merely repeating or "illustrating" what has gone before. In this case, *fiel* starts the movement downwards from the pinnacle of *glükt* to the ultimate abyss *dort*, where the poet, thrown *ins Dunkel*, will be lost to sight altogether. The reader cannot know this already, but the effect of *So fiel...* is to diminish or qualify the joy felt at the birth of song in the previous line.[49] When one is already in "*Glük*," as, for example, in the Garden of Eden, any "falling" is a fall "out of" happiness into misfortune (e.g., the *Sündenfall*). We notice also, once again, the "fall" out of the present into the past, away from the joyous completion back to a questionable, perhaps dangerous former state or event.

"...wie Dichter sagen..." The poet has been talking about poets and poems and now he reintroduces the "poets," but not for their own sake, as a phenomenon to be investigated, explained, or glorified, but as mediators of something more important than themselves. The story that the poets have transmitted—and that this poet transmits—is the focus. Actually, the poets themselves have been relegated to the shadows since St.II; at most they were the receptacle of divine production of song. Even so, it was their "soul" in which the miracle took place.

The narration of Semele's story has attracted much interest on the part of interpreters and with good reason, since it forms a crux of the entire poem. For present purposes, its importance lies in its form and its narrative temporality; its significance as content must wait for now. It must be noted at the outset, however, that this section of the manuscript is particularly difficult to decipher and leaves the impression of incomplete revision. The text as Beißner has re-constructed it has been attacked by many other interpreters, either in part or whole.[50] With that caveat, we will attempt to approach the text, such as it is. The narrative skeleton is roughly as follows:

1. *So fiel...* The ACTION: its tense shows it to have happened in the past, distanced from the reader. It occurred only once, as an example.

2. *...wie Dichter sagen...* The SOURCE of the story, the authorization of its truth. The question of truth remains open, however, since poets are widely assumed to lie.

3. *...sie...* The SUBJECT, which must lead at first to the false conclusion that it refers back to *Dichter*, instead of forward to Semele. What is interesting is that it does have a bearing on the plight of the present poet, which, however, does not appear openly until St.VIII (*"Doch weh mir!"*).

4. *...da sie...begehrte* These words give the REASON or explanation for something that remains unknown, although it has already occurred (cf. 1 above). The god in question is also unknown; he would seem to be one of the *Götter* whose collaboration with men made song possible (St.VI,4). He remains unnamed throughout the poem. Thus, what he does is disconnected from his identity.

5. *...sein Bliz auf Semeles Haus* The ACCOMPLISHMENT of the "falling" with which the statement began. The subject and the object are in the closest possible proximity to each other. *sein* refers back to *Gott*, *Bliz* back to *fiel*, *Semele* back to *sie* and also forward to *Haus*. We also cannot help noticing the frequent occurrence of the phoneme /s/—a visual and aural representation of the lightning, which also connects Semele's desire with its fulfillment by infusing "lightning" with Semele's name. By letting the name "Semele" appear (again only once), the poet at last clues the reader in on whose story is being told. Nor should the fact be overlooked that not Semele but her house is being struck. This is (with the exception *Lied* and *Waffenklang*) the only specifically human product in the whole poem. Similarly, it is not Zeus nor even *der Gott* who has cast his bolt; rather, it "fell," which clearly is an independent action requiring no actor. Even a momentary lapse of attention would suffice for it to drop. Thus, both Zeus and Semele are disconnected (removed, insulated) from their own story.

6. *die göttlichgetroffene...* (*"Getroffen"* = 1. decided upon [e.g., law, plan]; 2. captured [likeness, image]; 3. abashed, struck) The religious and mystical overtones are unmistakeable. Getroffen can mean that Semele was "touched" or "struck" (as in "it struck me that..."). Normally, one would assume such a person to be uplifted, perhaps ecstatic, especially in having a vision of the divine. Only when supplementary and text-external information is considered does the destructive aspect emerge.

7. *Und die ... gebar* RESULTING ACTION. Not even a comma interrupts the action; nor is Semele "killed." Rather than destruction, she merely loses her name, but also brings forth new life, which receives the main emphasis in this verse. She is described with a word that could be used in saints' lives or in the writings of the mystics. Thus the connection with *vom heiligen Stral entzündet*. The shift in gender also indicates that it was indeed Semele, and not just her house, who was intended above.

8. *Die Frucht ... Bacchus* RESULT OF RESULTING ACTION. The form reproduces the gradual coming to light of the newborn, achieving his identity at last. Syntactically, die Frucht could apply as well to Semele; semantically, though, this does not make sense.

If we re-late (*wieder-geben*) the story of Semele as it appears in the text, we find the interest of the poet focused on the lightning falling, on Semele as an individual person and her desire to see the god, and on her giving birth through this act of desiring/seeing to the "holy Bacchus." It is important to realize that the story does not so much describe a tragic destruction as a miraculous encounter with the divine which leaves in its wake the divine Bacchus. Taking into account the form of the narration, the suddenness of the event is stressed; it occurs before one knows what has happened.[51]

By now, though, the reader probably has a certain sense of *déjà vu*. The *Landmann* (St.I), the trees (St.II), the "I" (St.III), the "souls of the poets" (St.IV), have already experienced similar exposure to divine presence, with differences, to be sure—primarily in that they were not also exposed to divine destructiveness. Specifically, the description of the genesis of song (St.V–VI,5) has already introduced the essential motifs of Semele's story: the passive female subject is struck by an infinite being, "ignited" by his fire, and gives birth to a human/divine work—song—whose patron is Bacchus, and who also takes care of "*Glük*." The comparison also sheds light on aspects of Semele's story that are not explicitly stated. That she "shudders with remembrance," for example, or that Bacchus' purpose is to give evidence of his human/divine parentage, is not spoken of within the frame of Semele's story, but immediately preceeds it. This indicates that the point of the story does not lie with Semele's experience, but with the production of song and of Bacchus. The fact that not Semele but her *house* is struck by the lightning-bolt is consistent with this perspective on the *result* rather than subjectivity.[52] But what is the relation of *des Gemeinsamen Geistes Gedanken* to Semele or Zeus? They refer to the *Wettern* which wander through time and between heaven and earth and which also bring lightning. In other words, they are a manifestation of Zeus transformed into *Geist*, only now, instead of *Hinwandeln*, they strike *blitzartig*.

The major difference between Semele and the previously mentioned encounters with the divine (aside from the consequences) is that Semele "desired to see," whereas the poets were merely "exposed" to divine presence.[53] The *Landmann* went out to see, but apparently *after* the divine had already receded, as signified by the spatially distant thunder and the temporally distant lightning. Semele desired to see *sichtbar*, i.e., in the present, before her very eyes. In this, however, she was similar to the poets who are not impeded by any particular master, who instead are directly connected to Nature "herself" (*sie selbst*).

The second major question arises from poetry's dual nature. If song is *der Götter und Menschen Werk*, what does this actually mean? What part do the poets play that the gods could not, if they chose? Or can the two be separated? Perhaps this phrase only says that both are necessary to the production of song. The divine seems to draw on *Begeisterung*, as the source of the desire to create, the *"Feuer im Aug"* or *"...in Seelen der Dichter."* The poet brings the *Stoff*, language, in which the divine takes shape, just as Bacchus had a human shape, as well as divine abilities of metamorphosis.

The Third Triad (St.VII–IX)

After this recounting of a story from the past and *as* past, the poet returns in St.VII to the present (NOW). The first two lines demonstrate the ultimate result of Semele's story. Now men are able to drink *himmlisches Feuer* safely. This should surprise us, since the Semele story as usually told is rather a warning than a reassurance: it says, *"Stay away from direct contact with the divine,"* rather than *"Indirect contact with the divine is safe."* But, as we saw, the "contact with the divine" was not portrayed here as destructive; rather, it was portrayed as ecstatic, allbeit shattering, which corresponds on a naturalistic level to a sexual relationship which reaches through ultimate intimacy to "fruition." It should also surprise us that danger has been removed, since previously in the poem there has been *no explicit mention* of danger. Obviously, the Semele story has brought out the threatening aspects of exposure to the divine. It seems to have done so unintentionally, since, as we have seen, the mood of the whole poem was uniformly affirmative up to its introduction.[54] Even its manner of narration has worked to downplay its destructiveness.

The *"himmlisches Feuer"* is also ambiguous, and appropriately so. Bacchus, after all, is the god of wine *and* poetry (and song and ecstasy...). So, do mortals now *drink* wine without danger? Possibly, but certainly not in an absolute sense. Wine remains a "mixed blessing," even as the designation of "heavenly/fire" indicates. On the other hand, if *trinken* is taken metaphorically, then the poem is saying that we can now gain access to the divine through song. Such an interpretation is consonant with the description of the song's genesis in

St.VI, with *"im Liede wehet ihr Geist,"* (St.V,1), and with other statements in the text. The poet seems to be saying that, due to Semele's experience with Zeus, now (and henceforth) the divine is accessible to humans through poetry, and that it will not hurt them. We must ask, however, what do men "get" in poetry? If it is divine, then how can it be hidden, since its nature is infinite? That is, even if "hidden in song," the divine must still have contact with men—the same problem facing the poets, only deferred. Semele's action, which to the ancients was *hybris*, no matter how extenuating the circumstances, has become a self-sacrifice on the behalf of future generations of humans, similar to that of Prometheus or Christ.

But what is the nature of this fire? Previously in the text, fire has appeared as the sign of enthusiasm and inspiration in the poet, evidence of the presence of the god within. In such a state the poet could create song, or a man could "cast up high things." Fire thus provokes to action and creation and can possibly lead to improper action (*hamartia*). *Himmlisches* applies to the origin of this fire, but it could also refer to its goal or to the reaction it produces within men. In that sense, it could still harbor danger, even though this possibility is denied by the poet.

Up until this point there has been no clear differentiation between "human beings" and "poets." St.IV treats them as parallel, the *Landmann* could be either, the *Wetter* (St.V) are described in their relation to "us" without specifying "poets," and thus implying "us mortals." Now, however, a crucial distinction is made, introduced by the *Doch* of St.VII,3. The poet does not say that mortals are forbidden direct contact with gods, although the implicaton of Semele's story is that a mortal would not be "without danger" in their presence. He does say, though, that poets are supposed to expose themselves to the god in his storms *mit entblößtem Haupte*. It is thus the poets who are "naked" (*entblößt*), like Semele. Yet it is *they* who are threatened from the hidden. And indeed such seems to be the import of St.III, in which the poet relates his own experience of such presence. (Whether he experienced "presence" or merely "approach" is beside the point here, since even the latter is a kind of presence.)

This seems to be a rather strong affirmation of the poet's role in things, and in fact, in view of Semele and Zeus, more of a theological or priestly role than a "poetic" one. Momme Mommsen claims the poets were priests all along, but that the poetic voice only acknowledged the danger at the end (with the *"falschen Priester"*).[55] Previously, the poets were sheltered and raised by "divinely-inspired" Nature. Now they are "appointed" to approach the divine and precisely in its most dangerous form. Having just declared that the danger was past, thanks to Semele, the poet has drawn a new distinction which has then opened the door to new danger; whereupon he loudly and deliberately declares it the right and duty of all poets—and he calls attention to the fact that he is one of them (*uns*)—to march through it. He seems to go out of his way to court danger and blasphemy: *"uns gebührt es," "unter Gottes Gewittern," "mit entblößtem Haupte," "Des Vaters Stral.../ Zu fassen," "ihn selbst," "mit eigener Hand"* are just some of the expressions indicating his audacity.

Nevertheless, what the poet takes with his own hand is a "heavenly gift" and is destined for the "people." Whether this is to have a nationalistic tinge or not is hard to say, since the counterpart of *Volk* above was *Erdensöhne*, who would presumably include all men (and women?). On the other hand, clearly the receptivity of people to the poets' song is dependent on their sharing a common language, which would point in the direction of a particular people for a particular poet, though no significance is placed on which specific group is meant (e.g., Greeks, Jews, Germans). The point is that it is not from egocentric motives that the poets approach the gods to serve as channels; rather, they essentially negate their own identity in order to do so.

More interesting still is the gradual emergence of one single (though unnamed) masculine god in the course of the poem. Initially there was no "god," only *göttlichschöne Natur*, who was over all gods. In fact, her most outstanding trait was that she was *not* a a single identifiable entity, and that the poets, through allegiance or dependence on her, were not submitted to the rule of any (masculine) "sole master." Next, the *Kräfte der Götter* appeared, though in *Knechtesgestalt*, i.e., willingly subordinated to human needs. Then, after they were "recognized," we hear of the single *gemeinsam(er) Geist*, whose thoughts are responsible for planting the seed of song (not in so many words) in the now feminine and singular *Seele des Dichters*. Thus, a progression towards a singular, masculine deity, and a singular, masculine poet who takes on a feminine-receptive role (namely, as *die Seele des Dichters*) is plainly seen.

When the god's *Stral* is introduced, Semele's story is already prefigured, where this whole divine-poetic complex is given a decisive turn. With the god's firmly established masculine identity, augmented (twice) here by making him "God the Father" (*Des* Vaters *Stral*), things seem to get more dangerous for the poet, who in turn emerges from his anonymity and assumes the form of the poetic voice itself.

In the face of the impending confrontation of father and son, the poet reverts to childlike innocence, claiming that this is the only way poets can withstand the divine-paternal presence. I think the Christian overtones (*"Unless you become as children..."* [Matt. 18:3]) here are misleading. Aside from the possible sacrificial element in Semele's destruction, I see no reason for bringing in the whole nexus of Christian theological problematics dealing with sin/guilt/atonement and the Trinity. Rather, *vice versa*, what connects Christianity with the world of this poet is their shared prohibition against approaching the face to face.

The general thrust of St.VII, though, should by now be apparent. The poet is at pains to ensure that no danger accrue to the poets, and his defensive measures become ever more desperate. He says, *"Denn sind* nur *reinen Herzens...schuldlos"*—but of course what modern (i.e., post-Augustinian) person could honestly claim to fulfill such conditions? What is presented as a minimum requirement, easily met by most people, is in fact impossible without some form of massive repression or self-castration. And this requirement is presented *retroactively*, as a condition to protect against a danger which, although apparently always present, until now the text has not acknowledged.

In St.VIII the distance between high and low, between *Aether* and *Abgrund*, appears again, only this time (the only time in the poem) it starts with the low and brings the high near to it. The emphasis in these lines (St.VIII,1–4) is on suffering, the activity: *"versengt," "tief-erschüttert," "Mitleidend," "Leiden des Stärkeren," "hochherstürzenden Stürmen..."*. The heart, subject of the second clause, does nothing—it merely *remains*, endures through time. Although that may be sufficient, the heart's remaining—like the soul's in St.V–VI—is fundamentally passive. In St.VII the poets, as exceptional beings, approached the divine from below (hence *mit entblößtem Haupte, unter Gottes Gewittern*) to reach for the father's *Stral* and hand it "around" (democratically) or "downwards" (hierarchically?) to men. Now this same *Stral* does not crash down on the poets; nor has the poet ever explicitly stated that it would. Semele's story, however, has created the preconditions for such an assumption. Nevertheless, the poets are mysteriously tief*erschüttert*, they share (on the same level) the sufferings of the "stronger one," and as his hoch*herstürzenden Stürmen* approach, the poets' courage (symbolized as his "heart") "stands fast."

It should be pointed out how the syntax, starting from St.VII,7, bears the conditions and elaborations of the idea that "we poets are safe;" and yet, the effect of these lines is to increase the sense of danger, the immediacy of danger for the poet. Not only does he have to withstand the force of the god's storms, including thunderbolts; he must also in some fashion share with the god the latter's own suffering. I interpret *die Leiden des Stärkeren/Mitleidend* as conveying a *pathos* in the Greek sense of "something that happens to a body" (which may of course include pleasure or pain[56]).

The sentence ends with a dogged assertion that all can still turn out for the best. But the very next line begins: *"Doch weh mir/Wenn von..."*. It seems that even withstanding the storm of the last few lines was not enough to settle the storms the poet must endure. He continues to come up with new and (presumably) more terrible consequences, of unnamed but unfulfilled conditions, and this time the focus has shifted entirely onto the poet himself, as an isolated individual. We do not find out what might be able to break through the *vestes Herz* and cause such despair.

(A tempting source would be the *Prosaentwurf*, in which the poet speaks of a *"selbgeschlagener Wunde."* Although Peter Szondi has argued for its importance in clarifying this passage,[57] it must not be forgotten that precisely this part of the *Prosaentwurf* was *left out* of the metrical version. The problem would thus seem to remain.)

"Und sag ich gleich..." Momme Mommsen underlines the conditionality of this expression.

> Hölderlin sagt nicht ohne weiteres, er sei ein falscher Priester. Das konditionale "wenn" beherrscht die ganze Aussage . . . Das Wort "gleich" deutet hier auf einen hypothetischen Fall, der eintreten könne, aber nicht eintreten möge.[58]

Although in a conditional, hypothetical mood, the poet does indeed say—NOW, in this poem. It is tempting to conclude that the possible consequences he has projected are thus to be visited on him. The content of what he might say, however, is of little importance—he is still to be cast out.

Who throws him out? Notice that in *"die Himmlischen zu schauen,/ Sie selbst, sie werfen mich...," "Sie selbst"* has a dual function: as subject of *werfen*, but also as direct object of *schauen*. They are also thereby connected with *"Des Vaters Stral, ihn selbst."* Who are "the Heavenly Ones?" Presumably they are the gods of mythology, who have for so long tended mankind's fields in lowly forms. Now they cast him out, *den falschen Priester*, into darkness. This is the first we have heard anything about "priests"—what are their duties, who appoints them, how are they trained? On reflection, answers to these questions could be attempted, based on what the poet has already suggested, but the point here is that the poet is accusing himself of a crime—disloyalty to the gods—and implying that his view of himself is shared by the gods, without having so much as hinted at what has brought about this criminality. It seems almost irresistible to speculate about the nature of the poet's delict. M. Mommsen,[59] P. Szondi,[60] A. Seifert,[61] to name but a few, present differing and mutually incompatible interpretations, which, however, must remain speculation. He has presented this hymn from the outset as a more or less "neutral" party, or at least as one who has remained conscientiously in the background of things. Now he has stepped forward, no longer even in hypothetical form, to proclaim his own guilt and falseness. What are we, the readers, to think?

If the poet who has laid out before us in his poem the nature of poetry and of Nature herself turns out to be "false," i.e., unreliable, then he throws his entire text into self-contradiction. It does not seem possible that such self-contradiction was intentional: the poem does not seem to be "playing" with its own function as text, *à la* post-structuralism. Although the poet himself is consigned a monitory role, like the prophets of old, he is—like them—unlikely to effect a change in future hearers. Like Tiresias or Cassandra, he serves more to underline the tragic lot of men fated to live in awareness of their own exile from the gods and from ultimate truth. It seems thus that the poet is quite serious in what he says in the final lines.

Rolf Peter Norgaard, for example, sees clearly the devastating effect of the contradictions in the poet's own speech in view of the poem's concern with the *"transmission of the word."*[62] Whereas the *Landmann* went out to see the field on a *Feiertag* and saw the wonders of Nature returning to their rightful and life-supporting bounds; and whereas the poets, like Semele, have seen the divine in its splendor and, despite the tumult, have brought forth their song, which enables all men to take part safely in the divine; the present poet himself, who should above all incorporate the essence of poetry in his own poem, unmasks himself at its end as a Judas.[63]

But if this is truly the case, then the value of all preceding verses must be radically reinterpreted, even though all basis for such a revaluation is lacking. Either we must accept the description of Nature and the poets as it was presented in St.I–VII, in which case the poet is merely getting carried away in the last verses (which are unfinished anyway), or we must reject everything the poet has said, because he, as a "false priest," has no way of knowing all these things of which he has spoken. Even if he did, we could not rely on his veracity.

Or did he only become a false priest in the process of writing the poem? In other words, the poet began truthfully, but discovered in the process of writing that the truthfulness he was attempting was beyond his own powers and thus that the presumptions of the poem until then had been "false"—i.e., misleading. Was the poem itself, then, the divine presence which was to be withstood with *vestem Herzen*, so that the song could "succeed?"[64] *"Des Vaters Stral"* would thus refer rather concretely to the writing of this text. Or, instead, is the present text the consequence of the failure to withstand that presence? Norgaard writes,

> *"Wie wenn am Feiertage..."* is, in short, that very song that the poet sings as the false priest hurled deep beneath the living, the warning song that he sings to those willing to learn.[65]

The problems raised by this ending and its fragmentation go beyond the bounds of this initial commentary. It seems that, if we follow the words of the text, even if we include the syntax and structure of the poem, we are still left in a quandary. I do not believe this results from the subject matter, nor from the biography of the poet. Rather, it resides in the temporality of language itself and the nature of what the poet says, which forces the poet against his will to acknowledge *ἓν διαφέρον ἑαυτῷ*.

Many interpreters have attempted to find a general "message" in *"Wie wenn am Feiertage..."*. Any such message, though, would be subject to the limitations placed upon it by the language in which it appeared. If we wish to speak about time or temporality, we must of course employ language. But language itself first enables a recognition of temporality through its *sinnstiftende* function. All language employs tense and deixis in order to create a "scene of utterance" which is privileged (taken as "real") over all others. The text, by creating a "place" for language to occur, also creates a sense of time as a way of indicating something which either does or does not belong to that place. G. Scimonello underlines the importance of Hölderlin's conception of time when attempting to abstract his "thought" from his poetic "work."

> Während Schelling und Hegel die Zeit entweder aus dem undifferenzierten Absoluten oder aus der Idee ableiten, *läßt Hölderlin das Absolute in die Zeit aufgehen*. Folglich existiert kein Absolutes für den Dichter..., sondern nur noch das zeitweilige Übergewicht einer Macht, die sich zur Totalität erhebt, indem sie jedoch ihre zeitliche Struktur beibehält.[66] [italics added]

But the time-space created by language is unstable. It may appear perfectly clear, delimited from all other places, but through time it shows itself to be changeable. Impermanence is destructive to the very idea of stability, identity, truth, etc. An awareness of impermanence threatens to make any type of action impossible, for all the components and even the action itself—and even we ourselves as unified subjects—are thereby put into question.

Hölderlin opens the door to a new awareness of temporality in *"Wie wenn am Feiertage..."* He does not leave it open, though. It is a notion of "being" which includes "becoming" and "passing-away."[67] As such, it is self-contradictory, self-destructive. If being is only to be glimpsed in becoming, then any attempt at "fixing" it in conceptual language must fail. It may be hinted at, it may be brought to light; but it is unmistakeably also a process, a movement which takes "time" (in the ordinary sense). For this reason I have concentrated on the "how" of the poem, as much as on its "what."

The difficulties which this poem has presented to our analysis can be traced, I believe, to the methodology we pursued. The text has been approached with a view to uncovering its temporality through evidence provided by linguistic signs in the text and by the poetic structures that provide context for them. Additionally, temporal implications of individual words and phrases have been brought out for consideration.

We have thus followed a method based on the FORM and CONTENT of the text, taken as linguistic artifact.

The Introduction to the study, however, indicated that we would be following a phenomenological-hermeneutic approach. It may have seemed that such has been the case. The initial approach discovered problems which a more thorough reading was designed to resolve, or at least systematize. The first step in the hermeneutic process has been taken, but it seems to have led astray. The poem has been "analyzed"—taken apart—and the manifold aspects of temporality at work in it have been brought out, but it does not stand as a "text" anymore. It now must be restored to its original integrity.

The structural-linguistic approach has indeed uncovered "temporality" in the poem, but the kinds of temporality it found were the kinds it was seeking. The poem has become in this way an excavation site for temporal "evidence," but has ceased to live as poetic speech. The poem's difficulty has to some extent been understood, but its source of attractiveness to generations of readers remains obscure.

I believe *"Wie wenn am Feiertage..."* has still more to tell us about time. To hear that message, though, requires a new way of access to the poem, one that does not damage its own way of saying. The hermeneutic circle must turn again, this time to find the poem in the right way. To do this, we will turn to three texts by Martin Heidegger. Through Heidegger's reflections on art, we will at first turn away from this specific poem and its problems, to focus on the broader issue of the "origin of the artwork." Then, after learning through Heidegger what the artwork is, we shall return and turn the light of our investigation on his own version of *"Wie wenn am Feiertage...."*

Chapter Three

"Der Ursprung des Kunstwerkes": HOW TO FIND AN ARTWORK

Unlearning Form and Content

THE INTRODUCTION TO THIS STUDY treated temporality under the double headings of form and content. The initial reading of *"Wie wenn am Feiertage..."* (Chapter Two) showed the limitations of such bifurcation when reading a literary text. Heidegger, in *"Der Ursprung des Kunstwerkes,"* not only disputes the usefulness of concepts like "form" and "content," but shows how they actually bar the way to an appropriate understanding of works of art, including, of course, literary works of art. In order to understand temporality in *"Wie wenn am Feiertage...,"* we must know how to approach it. The way chosen in Chapter Two led into confusion; the temporality of the poem exceeded its reach. We must now step back and re-approach the poem in its own way. *"Der Ursprung des Kunstwerkes"* suggests how to do just that.

What is most important about the work of art, according to Heidegger, is the way it discloses things as they are "in truth," without the everyday trappings that usually trap them in invisibility. "Beauty" in art is important but it is not an irreducible element. Beauty (*Schönheit*) is the appearance (*Scheinen*) of this truth in the work, but not in the sense of an "effect" of some ideal cause or the signifier of an absent signified. The "truth" which "goes to work" in the artwork is neither empirical nor rational. It is not a fact and not a concept. It cannot be discovered by collecting data or by arguing from first principles. It is not an aesthetic judgement. When the artwork does its "work," it places an historical humanity (*Menschentum*) before its fundamental decisions (*Entscheidungen*), i.e., the answers to questions which will determine what truth is for that humanity. *We* do not judge the artwork; the truth of the artwork judges *us*.

"Der Ursprung des Kunstwerkes" ultimately claims pride of place for *Dichtung,* as the most fundamental way an artwork works. *Dichtung* is usually translated as "poetry," but in German it is a broader term, including all forms of "literature," taken as *belles lettres*.[1] To avoid confusion, since there is no adequate English term for Heidegger's use of it, I have left it untranslated. Heidegger uses *Dichtung* to mark the special relationship language has to man: it is through and in language that man lives.

In order to bring out the working of temporality in the poem, Chapter Two looked for (structural) temporal markers: rhythm, repetition, tense, and aspect, on the one hand; and words having temporal content, like *"jezt," "die ganze Zeit,"* and *"der Feiertag"* on the other. We found a basic contradiction between the time as announced initially (*"jezt aber tagts!"*) and the time at the poem's end. This split in the temporality of the poem could be interpreted as stemming from the text's own fragmentariness. If that were true, the poem would simply be a "failure," an unsuccessful first attempt at what was for Hölderlin a new form of poetry, the Pindaric hymn. Perhaps, though, the problem with the poem comes from the kind of temporality we have been seeking. Might it be that the poem's temporality has not even yet been found, that we have simply *imposed* a temporality on it?

In *"Der Ursprung des Kunstwerkes"* Heidegger unfolds a new way of looking at works of art. He sets aside the traditional schema of "form" and "content." He does not, however, argue against that schema directly. Instead, he asks what the proper way to find the artwork would be. Only by seeking the artwork in its own way can the danger of imposing inappropriate expectations on it be avoided. We, too, must heed that warning and approach Heidegger's own text on its own terms. Clearly it differs from most considerations of art, whether ancient or modern. Its divergencies, however, are not the signs of failure or ignorance, but are attempts to re-think the artwork. Our own reading of *"Der Ursprung des Kunstwerkes"* also seeks to understand the aesthetic process better. Where do we find an artwork?

Using temporality as a guide, we will first draw out Heidegger's method of approaching the artwork. It will immediately be clear that it is a phenomenological hermeneutic as described in Chapter One. Heidegger begins with common, everyday formulations of the question of the artwork and, using the phenomenological method, demonstrates the difficulties hidden within that terminology. Robert Mugerauer, discussing another work by Heidegger, writes:

> Though Heidegger does in fact tell us about thinking, his primary concern is to exhibit thinking; the "end" of the address is not so much to tell us this or that, as it is to engage us in the process of thinking.[2]

Heidegger's hermeneutic necessarily traverses the rough path of error and illusion, of cliché and wild goose chase (*Holzweg*). Such is his "process of thinking." Having discovered a circularity, contradiction, or undecidability within the hermeneutic situation, the investigation returns to the initial formulation of the problem and re-thinks it. The return reveals a new view of the matter and leads to more sustained inquiry. This is how the questioning of the artwork proceeds.

After detailing the phenomenological method used here, we will consider how it transforms the traditional concept of the artwork. The poetic text ceases to be a kind of "thing:" beautiful perhaps, but ultimately distinct from its reader and at his or her disposal. Instead, the work of art will be shown to be where the conflict of "world" and "earth" takes place. The world of the *art*work opens possibilities of saying which the earth of the art*work* reduces to a single, though "real," one. The work in turn makes it possible for man to come before the essential decisions of his history. History is conceived here as not merely a succession of moments, dates, ages, etc., but as a kind of time in which truly decisive turns are made. What makes a turn decisive is that it cuts off the possibility of *turning back* to a previous time.

The kind of artwork which most fundamentally brings man before such decisions is *Dichtung*. *Dichtung* is the essence of the artwork, according to Heidegger, because it is language in its original role of disclosing the truth of beings. Having presented *Dichtung* as the foremost art, he has prepared the way for further investigations in *"Hölderlin und das Wesen der Dichtung"* and *"Hölderlins Hymne 'Wie wenn am Feiertage....'"*

Heidegger's Method

This work of 1935/36, which originally appeared as a series of three lectures, given several times, expanded, and ultimately published as a unit in *Holzwege*,[3] is important for demarcating the field and method of investigation to be followed in the present study of *"Wie wenn am Feiertage..."*. Our own reading of Heidegger's essay will begin by remarking that it shows clearly his typical approach. Although he does not explicitly state it, he seems to be employing a phenomenological-hermeneutic method similar to that of *Sein und Zeit*.[4] He first sets out the problem in naïve, commonsense terms and then easily shows their own difficulties; these, however, lead back to a better understanding of the original problem, enabling a more serious approach to an answer. In this case, the topic is announced by the title, *"Der Ursprung des Kunstwerkes."* Where is the "origin" of the artwork to be found? The origin of the artwork is, *"according to the usual conception,"* the activity of the artist. But the artist himself only becomes what he is through the artwork. Neither can claim priority over the other; in fact, each relies upon something else for its very existence: namely, "art."

Already the outlines of Heidegger's method can be seen. First, the problem or topic to be discussed is presented naïvely, in sometimes provocatively simple words. Using only common understandings of the words through which the problem is presented, he then quickly shows their inadequacy or (as here) circularity. The artwork derives essentially from the artist; but the artist is only truly known by the artworks he produces. This rather Socratic gambit throws open the discussion to a new attempt at understanding; one which of necessity must be somewhat more thoughtful, i.e., less obvious. As Gerald Bruns puts it,

> One reads the later Heidegger, not in order to do something to him or to get something from him, but just because the questions of language, truth, and

> poetry that he raises (as does no one else) remain endlessly, maddeningly, open. They evade the grasp of our concepts.[5]

The basis for both artwork and artist is "art," but what does this mean? "Art" is, after all, only a word. How can a word be anything other than an empty term, relying for its being on its "real" referents, viz., the artist or artwork? In other words, "art" does not build cathedrals or paint pictures; only "artists" do. And no one who does not at some point come forth with a "work" can truly call himself an artist. Heidegger does not dispute or disprove such a conception. Rather, more subtly, he *suggests* an opposite way of looking at the matter: could it be that artist and artwork can only be insofar as art is? The question functions here as a heuristic means out of a quandary. Heidegger is not yet in a position to *demonstrate* the priority of art over artist and artwork, but since the possibility is offered *gratis*, as it were, it is difficult to refuse.

He does not simply disregard or minimalize the evident circularity in his statement of the problem. Rather, he first offers an empirical and then a deductive (rationalist) alternative formulation, but quickly exposes their inadequacy: each proves to be based on presumptions which cannot reach the "essence" of the artwork.

> Das Aufsammeln von Merkmalen aber aus Vorhandenem und das Ableiten aus Grundsätzen sind hier in gleicher Weise unmöglich und, wo sie geübt werden, eine Selbsttäuschung. [UK,2]

He therefore concludes, rightly, that the investigation cannot break or avoid this circularity, but must instead follow it to its end (*"So müssen wir den Kreisgang vollziehen"*).

The path begins with the phenomenological certainty possessed by each of us that he or she can recognize a "real" artwork if presented with one (*"I don't know much about art, but I know* real *art when I see it"*). But if we look for "real" artworks, we consider them as "things," no different from, say, hunting equipment or hats, both of which also hang on walls. If we take our search for the "origin" or "essence" of the artwork to mean a search for the "real" artwork, then it can only be found in "reality," which, as everyone knows, is composed of "things."

Certainly the conception of reality in the physical sciences, based on theories of matter and energy, sees the world as made up of "things."

No one, however, would likely be satisfied with an understanding of the artwork as merely one kind of "thing" among many other kinds. Even if art is in some way "thinglike," it is also responsible for an experience which is more meaningful to some than to others. That is, if we are going to select artworks on the basis of consensus (what "everyone" knows), we must also consider carefully who is eligible to vote: *"Wir müssen doch die Werke so nehmen, wie sie denjenigen begegnen, die sie erleben und genießen."* [UK,3] Some people seem to be better qualified to pass judgement on whether such-and-such a thing is an artwork or not, and if so, how "good" it is. They "experience" (*erleben*) more and articulate their experiences better than others. And, since they are able to articulate their opinions better than others, their judgements have a weight (*Tragweite*) lacking in those of others.

The artwork, in expressing something "else" (*Anderes*) than mere thingliness (the straightforward, literal meaning of words, or the woodness of wood, or the stoniness of stone), is also engaging in allegory (*ἀλλο−ἀγορεύειν* = to say something "else" [than the public meaning]). "Bringing together" (*συμβάλλειν*) the other things being said in the artwork makes it a "symbol" as well. The simpleminded view of artworks as "things" thus gives rise to the basic concepts of aesthetics, which have dominated discussion of art for centuries.

After laying out all these aspects of the artwork, Heidegger insists that they themselves are still problematical. Before their value for a discussion of the artwork can be determined, their own dependence on a more basic concept needs to be understood. The concepts "experience," "allegory," and "symbol" all derive from an understanding of the artwork as a kind of "thing." That is, the experience is "of" something, the allegory refers to some "other" thing, the symbol symbolizes something. It is therefore incumbent on us to consider more carefully the nature of the thing (*das Dinghafte*). Only after that is done can it be determined whether the artwork is a thing after all ... or not.

All the traditional concepts of aesthetics have been placed in question. The way is thus open for a new look at what an artwork is. At this point a résumé of Heidegger's process might be helpful in determining the way to go.

1. A question is raised in simple, seemingly unproblematic terms.
2. A fundamental contradiction or circularity appears.
3. A resolution to complete the circle is made.
4. A further effort to understand the topic naïvely is made.
5. Some fundamental concepts of traditional aesthetics are derived from this "naïve" understanding.
6. A need is shown for a return to the initial starting place (here: the "thing").

This cyclical process can be found in many of Heidegger's works. It exemplifies the phenomenological hermeneutics described in the introductory chapter of this study (pp. 12f). It begins with the most simple and straightforward statement of the issue and proceeds to deepen and transform it by repeated returns to the initial formulation. Thus, we come to *hear* those words differently and to understand the problem differently—not by a chain of reasoning which leads us away from the words—but by a way of becoming *hörender*, more sensitive, to those very words. The question itself becomes more questionable. Finally, a hint is given that this path is actually a (necessary) detour through error.

The temporality of *"Der Ursprung des Kunstwerkes"* is cyclical on a small scale through Heidegger's use of the "hermeneutic circle." It is cyclical on a larger scale through the sustained effort of thought which courses throughout the lecture series. It progressively wears away the easy, apparently "right" concepts, starting with the artwork as thing. This process does not go on indefinitely; it leads to the decision which we, as hearers or readers of Heidegger's text, face (...have faced, will face). The decision puts an end and a beginning to the hermeneutic process. As we shall see, that process takes place as *Dichtung*.

The larger circle also begins with the initial statement of the problem: what is the origin of the artwork? Heidegger's answer to that question traverses the entire lecture series. It begins with crude if self-confident materialism and ends, having dissolved all partial solutions, by re-posing the question of the artwork. To do so, it passes through several stages, beginning with the question: What is the origin of the artwork?

1. The artwork is not a thing; in fact, things are only truly accessible through the artwork.

2. The artWORK sets the truth to WORK. (It is activating.)

3. The artwork sets up a world and sets it back onto an earth.

4. The conflict of world and earth in the artwork confronts a historical humanity with the decisions of its history, both as something "past," and as yet to be made.

5. The work relies on those who stand within its truth and thus are its "preservers." They take a stand with decisiveness on the world of the work.

6. All art is essentially *Dichtung*: a clearing which makes possible the decision.

7. Are we now historically at the origin? That is, how does it stand now with regard to the decision about the originality of the artwork?

Heidegger leads the reader on a process of approach and withdrawal, of disclosure and concealment. Taking his hint, we must follow this wavelike washing away of each layer of appearance to reach his "truth."

In the remaining sections of *"Der Ursprung des Kunstwerkes"* this process is often repeated. It leads from a consideration of the artwork as thing to the artwork as equipment, to an understanding of the artwork as bringing out the truth of beings, to a deeper understanding of truth and art. In each case, the initial understanding of the terms must recede before repeated questioning. Gradually, the artwork comes closer as itself. The next section of this chapter will look at each stage of this process.

The Path to the Artwork

In order to understand the nature of artworks we must take what they offer us, even if that be something as obvious as their "thinghood." Undoubtedly artworks, no matter how sublime, also exist as paper, stone, paint, sound, etc. "Real" artworks seem to exist just like everything else in the world. Beethoven's quartets occupy space in publishing warehouses just like "potatoes in the cellar," to use Heidegger's example. Is he about to fall into the materialist trap of describing "art" in purely empirical terms? No, but he is calling attention to a fact about artworks that is all-too-often passed over in silence in discussions of art. Artworks, grossly seen, are "things." But what exactly is a "thing?"

Things are the most common "thing" imaginable. But they are by no means the best understood. There seems to be different kinds of thing, like rocks and pitchers and milk and clouds and thistles. Even something that could never appear itself—like God, or the universe—is also some sort of a thing. But we seem to have difficulty calling animals and men "things." The true thing is the "mere" thing, stripped of any further attributes. Even as sense impression its thinghood is not reached; phenomenologically speaking, we do not perceive an inrush of impressions, rather *"we hear the storm whistling in the chimney and ... the Mercedes in direct distinction to the Adler."* [PLT, 26] Even taken as the unity of matter and form, the thing remains elusive. According to Heidegger, we go astray by trying to place the "thing" too near our immediate experience. To return to the simple thing we need to look at where the idea comes from in the first place.

Even if we exclude animate nature from the concept of "thing," we have still have not reduced it to its simplest form. Some things are closer to man than others: i.e., some things owe their very existence to man's activity. Accordingly, the cup or the fireplace would be closer to us than the stone in the field or clouds in the sky, because the former have been made by humans, whereas the latter seem to exist independently of humans. Heidegger calls things which come into existence through humans and for humans *Zeug* (equipment). He says that equipment, having been made by humans for their purposes, is characterized by its reliability and usefulness.

In order to get a closer look at the nature of equipment (*zum Zeughaften des Zeuges*), Heidegger takes an example: a pair of peasant's shoes. But how to approach them, such that their equipmentality will emerge clearly? It would be naïve to imagine that simply *placing* a pair of shoes before us would be sufficient, or that a straightforward photograph would do. Heidegger chooses instead a painting by Van Gogh which portrays a pair of shoes. Since it is obviously also an artwork, i.e., the focus of this entire discussion, it seems a particularly loaded choice. But Heidegger immediately discounts the significance of the painting as a means of representing what he is after. The discussion concerns the *origin* of the artwork, its essence; it is not after an exact description of a particular pair of shoes. In this case, it is the essence of thinghood that is being sought through the equipmentality of an "ordinary" pair of peasant's shoes.

> Wir wählen als Beispiel ein gewöhnliches Zeug: ein Paar Bauernschuhe. Zu deren Beschreibung bedarf es nicht einmal der Vorlage wirklicher Stücke dieser Art von Gebrauchszeug. Jedermann kennt sie. [UK,17]

Later in the same paragraph he calls the painting "well known" (*bekannt)* and says that "everyone knows" what constitutes a shoe (*"Jedermann weiß, was zum Schuh gehört"*). This emphasis on the everyday commonsense view of things is the marker of a renewed entry into the hermeneutic circle, of which the essay as a whole is the primary example. Shoes can be made out of wood or leather; they are stitched together with needle and thread; their usefulness consists in covering the feet; their form and material follow their function, which is "serviceability" (*Dienlichkeit*). These are all "well-known" facts about shoes and Heidegger does not dispute or disparage them. On the contrary, they point to true serviceability, which can only be found when the peasant woman *wears* them in the field. Not through "representation" can we truly appreciate the shoes' usefulness; only through using them.

It seems as though the painting itself as artwork has contributed nothing special to this discussion of equipmentality. It portrays the shoes so isolated from any realistic environment that one cannot even tell where they have been placed. And yet those shoes tell a story, they bring forth a world and an earth. What would the effect have been, if Heidegger had produced two actual shoes? We would not thereby hear their story better than in Van Gogh's painting. We probably would not have heard the story at all. The story of the peasant woman's shoes speaks only from within that painting; even the peasant woman herself does not hear it more clearly. Every day she puts on her shoes and through them has access to a world of fields and crops, of weather and concern about the harvest.

> Das Zeugsein des Zeuges besteht zwar in seiner Dienlichkeit. Aber diese selbst ruht in der Fülle eines wesentlichen Seins des Zeuges. Wir nennen es die Verläßlichkeit. [UK,19]

Fundamentally, the usefulness of equipment comes from its "usualness," its ability to bestow familiarity on the world and thus to let it disappear in self-evidence. The equipment gets worn away in its very familiarity until it is simply "there," to be used without thinking. It thus grants a kind of peace (*Ruhe*) to the world, since it is always the same dependable equipment. The car we use everyday to drive to work disappears in its reliability and even brings the outside environment to rest (since we always participate in that environment through the same means), thereby freeing us to daydream. Such is the essential nature of equipment, according to Heidegger: not a focus of our concerns, it *is* in its invisible serviceability. Has the nature of the artwork been discovered? The artwork undeniably shares something with equipment: it too grants access to a world, but in a contrary fashion. Rather than resting in invisibility, the artwork brings things forth with unaccustomed clarity and forcefulness.

Before launching on a further exploration, Heidegger summarizes the path just completed. This recapitulation serves to mark the transitional "end" of the hermeneutic circle of equipment, which began with a consideration of the nature of the "thing." The artwork is not a thing and cannot be understood properly if we seek it from this direction. This brings up the necessity of a reevaluation of the premises with which we began.

This short lecture illustrates once again the operation of the phenomenological hermeneutic as Heidegger practices it. The common and familiar was made to reveal much more than could be attained through strictly discursive means. The nature of equipment was made clearer by describing a particular and familiar example phenomenologically. It must be emphasized once again how critical the familiar and everyday are for Heidegger's method. Only they can give the discussion its foundation in the world; without them, it remains merely abstract and insubstantial reasoning. The naïve is not used as a convenient foil for the philosopher's more "sophisticated" thought, but as an equal partner in the search for an answer to the initial question. Heidegger did not bring in the peasant woman's shoes to have a little fun at her expense. Her relationship with her shoes is not less for lacking words to describe it. At the same time, it does not have the artwork's self-reflectiveness. Heidegger himself is fully aware of the importance of the naïve in understanding the artwork and the importance of the artwork in discovering the essence of equipment, which appeared so obvious before.

> Aber vor allem diente das Werk nicht, wie es zunächst scheinen möchte, lediglich zur besseren Veranschaulichung dessen, was ein Zeug ist. Vielmehr kommt erst durch das Werk und nur im Werk das Zeugsein des Zeuges eigens zu seinem Vorschein. [UK, 20]

The work lets the *Zeugsein* (equipmentality) of the shoes come into the open, it lets them appear as what they essentially are. By bringing beings out of their concealment in the everyday, the artwork—and it alone—is able to show them as they truly are. This kind of truth, the truth of unconcealedness, is what the Greeks called *aletheia* and what Heidegger translates as *Unverborgenheit*. Thus, the essence of art is: *"the setting-to-work of the truth of a being."* [UK,21]

Heidegger again recapitulates briefly the path he has taken up to this point in the discussion [UK,25]. Having sought the most immediate reality of the artwork in its thingliness, as "*geformter Stoff*," he has found the origin of that view in an inadequate concept of the thing as equipment which cannot reach the essence of thinghood. Although its thingliness is inadequate to understanding the essence of the work of art, it is not "mere" material to be ignored or overcome in the quest for a "higher" aesthetic reality. The artwork is and remains a kind of thing, and this is constitutive for it, without, however, being sufficient to understanding it. Rather than explaining the artwork through the concept of the thing, it is on the contrary only in the work of art that the thing is uncovered as what it truly "is," in its *Wesen*. Heidegger emphasizes the intransigence of the work as thing in the total working of the work of art. He says

> Das Dinghafte am Werk soll nicht weggeleugnet werden; aber dieses Dinghafte muß, wenn es schon zum Werksein des Werkes gehört, aus dem Werkhaften gedacht sein. [UK,24]

But what is this "workliness" and how does the truth go to work in it?

The topic of the second lecture has changed somewhat from that of the first lecture. We have seen how art becomes real in the work of art, and how the work is not comprehensible through its thingliness. The poem is not identical to its pages, nor is the painting identical to its paint. What must now be understood is what the work truly is, "in itself" (*an sich*). This is not so straightforward as it may at first seem. Poems are kept in libraries, paintings in museums; but are they there as themselves or as *"objects of the art business?"* Even if they were painted on commission, they never "belong" in a museum or a library. To say that they did would imply that they were fraudulent as works of art. We have seen how even peasant shoes carry their world with them; how much more so does the artwork? But the artwork is almost always encountered at second- or third-hand, through the efforts of its preservers. Nor can this preservationist element simply be avoided: "great" art has always already been torn from its world, where it came to be. It is always "handed over" (*über-liefert*) to us by a tradition (*Überlieferung*).

> Die Werke sind nicht mehr die, die sie waren. Sie selbst sind es zwar, die uns da begegnen, aber sie selbst sind die Gewesenen. Als die Gewesenen stehen sie uns im Bereich der Überlieferung entgegen. Fortan bleiben sie nur solche Gegenstände. [UK,26]

So the question as to how to find the artwork has reappeared. This was, in a sense, the opening question of the whole discussion. What is the "origin" of the artwork? What is its "descent" (*Wesensherkunft*)? The truth of the artwork, its bringing forth beings as they are in themselves, is thus *historical*. It is impossible to discuss artworks without including some reference to the tradition out of which they speak. The question is, rather, which tradition is that? Are they to be taken simply as historical artifacts (*Gegenstände*)? Heidegger answers that

> Das Werk gehört als Werk einzig in den Bereich, der durch es selbst eröffnet wird. [UK,26]
>
> The work belongs, as a work, uniquely within the realm that is opened up by itself. [PLT,41][6]

Heidegger here insists on the absolute autonomy of the artwork *vis-à-vis* the World. It resides in its own world, the only proper one for it. To avoid any suggestion that the world of the work imitates or represents the world taken as physical reality, he chooses to examine a non-representational work, a Greek temple.

As with the choice of Van Gogh's painting, the choice of a Greek temple has led to the mistaken assumption that Heidegger himself is promoting it as somehow a "superior" example of art, i.e., "great" art. As Kockelmans points out, Heidegger is following Hegel's practice of discussing works commonly recognized to be artworks.

> "Great art" refers to those works that were produced in Greece, Rome, and the Middle Ages, and which are such that there is today common agreement on their artistic status. Great art implies the totality of all art works, made before the time in which the fine arts came to the fore as such... [7]

Although he does not explicitly say so, this concept of "great art" is independent of culture. It does not depend solely on a Western perspective, though that is the one chosen by Heidegger in his address. We must remember that Heidegger is not doing cultural anthropology here, but looking at the nature of art. By starting with what is most familiar to his audience, namely Western art, he seeks to build on that understanding which each of us has but which is seldom scrutinized. The phenomenological hermeutic begins with the average everyday understanding of the artwork as the only suitable point of access. It does not exclude extensions to other, non-Western traditions, but is simply not able to proceed without first understanding its own foundations.

Heidegger now introduces two concepts which will be crucial for finding the artwork: the work sets up a "world" and sets it back on the "earth." "World" is the totality of meaningful relations which the work makes possible.

> Welt ist nie ein Gegenstand, der vor uns steht und angeschaut werden kann. Welt ist das immer Ungegenständliche, dem wir unterstehen, solange die Bahnen von Geburt und Tod, Segen und Fluch uns in das Sein entrückt halten. [UK,30]

Thus the world of the work is not something isolable and definitive, but it first makes possible the emergence of the work itself. The statue of the god in the temple is a work only as long as the god has not fled, as long as the world (of gods and men, of temples and sacrifices, etc.) still *is*. "Welt weltet *und ist seiender als das Greifbare und Vernehmbare* ..." [UK,30].

World is also not just the "form" which "matter" takes in a work. As opposed to equipment, which can be said to "use up" its material when it has been made, the work first brings out the essential nature of its constituents.

> ...der Fels kommt zum Tragen und Ruhen und wird so erst Fels; die Metalle kommen zum Blitzen und Schimmern, die Farben zum Leuchten, der Ton zum Klingen, das Wort zum Sagen. [UK,30]

But the work does not really "bring out" something from matter. Rather, it provides a place for things to come and in so doing it sets the world of the work back onto an earth. The work of art invites things to come forth and dwell with it for a while, as at a picnic, spreading out its offerings on a suitable plot of ground. The work is not just some collection of imaginary forms (pure ideas) with no substance or extension. The "world" of the work is *real*, in the everyday sense, and its "reality" depends on an "earth." Heidegger always insists on the "reality" which the "thingliness" of artworks preserves.

> Wenn wir die Werke auf ihre unangetastete Wirklichkeit hin ansehen, dann zeigt sich: Die Werke sind so natürlich vorhanden wie Dinge sonst auch.
>
> [UK, 3]

Of course, by "earth" Heidegger is not simply referring to a particular planet in the solar system, nor to a clump of dirt. Earth is that which "grounds" the world of the work and thus makes possible a place for historical man to dwell.

> Auf die Erde und in sie gründet der geschichtliche Mensch sein Wohnen in der Welt. Indem das Werk eine Welt aufstellt, stellt es die Erde her.
>
> [UK, 31–2]

Earth is that which announces itself by its withdrawal from every attempt to penetrate it. The stone is heavy and announces itself by being heavy; any attempt to understand its heaviness through concepts is shattered against its withdrawal into itself. According to Heidegger, it is *by* closing itself off that earth announces itself. In terms of poetry, the poet "uses" words, but does not use them up (consume them), as instructions are used up and thrown away after they have been followed. Rather, he brings them forward as what they truly are, and thus confronts the reader with their opacity. The poem sets up a world in which constituent words step forward as "earth."

World and earth always appear together, in what Heidegger calls the essential conflict (*Streit*). World always exceeds our grasp by refusing to be pinned down to a single entity. Earth always denies entry by withdrawing into itself. It is only through their combined working that each comes forth as itself. In that way, this conflict is essential because it goes to the essence [*Wesen*] of the relationship of world and earth. *"Im wesenhaften Streit ... heben die Streitenden das eine je das andere, in die Selbstbehauptung ihres Wesens."* [UK,34] The work of art starts up this conflict between world and earth and fights it through. Only in this fashion does it achieve its "unity," which is no mere "harmonizing" of disparate elements. The conflict of world and earth produces a kind of equality, a relationship of attraction and repulsion which each shares with the other. Heidegger's term for it is *der Riß,* not in the sense of "tearing apart," but as *"das einheitliche Gezüge von*

Aufriß und Grundriß, Durch- und Umriß." [UK,50] Gerald Bruns calls it the "common ground." [8] Although Heidegger's language here lends itself to martial misinterpretation, his notion of conflict must be carefully distinguished from mere fighting in a biological or sociological sense. As I read him, Heidegger is thinking conflict in terms of a heraclitean *eris*, which is the foundation and presumption for all things, including especially any *harmoniê*.[9]

Truth has many meanings. In the present instance, we are after the "essence" of truth, not just this or that truth. Post-cartesian philosophy has generally taken truth to mean "correctness" (*Richtigkeit*) in propositions. But all "correctness" already implies an openness of Being within which "conforming to something" (*sich richten nach*), as well as the purpose of such conformity (as a possibility of being), becomes manifest. The German words show clearly the function of propositional truth in giving a "direction" (*Richtung*) to man, helping him to orient himself.

Heidegger takes as name for this openness *aletheia*, the Greek word usually translated as "truth," but which, according to him, literally means "unconcealedness" (*Unverborgenheit*). *Aletheia* seems to involve just a switch of names, nothing special. Heidegger claims, though, that even the Greeks did not understand the unconcealedness of *aletheia* fully. As Heidegger uses the term, truth as unconcealedness always exists together with its Other, concealment. Just by disclosing itself, truth must hide the *way* in which it has come into the open. That is, for truth to be recognized as such, it must be taken as "absolute," "final," "certain," and yet, on reflection, it is clear that what is taken to be true has *changed over time*. Truth is never static; it is a fluctuating process of illumination, of coming forth into and retreating from unconcealedness, in which temporality plays a constitutive role. Christoph Jamme underlines the temporality of Heidegger's concept of truth as follows:

> Wahrheit wird hier bestimmt als Zusammenspiel und Widerstreit der Freiheit des Entwurfs und der Geworfenheit des Entwurfs, als Einheit von Unverborgenheit und Verbergung und damit als Prozeß.[10]

Religious, historical, even scientific "truths" do not remain eternally the same, although we necessarily act as though they do.

To say that something stands "unconcealed" implies that it is plainly in view. But the kind of unconcealedness Heidegger is talking about is not so easily described. Even though it is the most familiar thing in the world, it is difficult to conceive of, much less talk about. We come upon unconcealedness as a "clearing" (*Lichtung*) in which things can appear as such. The clearing, by making possible the appearance of beings, is more primordial than the beings.

> Inmitten des Seienden im Ganzen west eine offene Stelle. Eine Lichtung ist. Sie ist, vom Seienden her gedacht, seiender als das Seiende. [UK,38]

The clearing (*Lichtung*) in which beings stand is also a twofold hiding (*Verbergung*). It is a hiding in the sense of a refusal (*Versagen*) of everything except the one thing that this thing is; and it is a hiding in the sense of a disguising (*Verstellen*). Thus, truth as unconcealedness includes a twofold concealing. We see once again how contradiction and conflict (*Streit*) are essential for things to appear: in order for one thing to appear, it must refuse all else, which is necessarily a disguising of the whole. Further, truth is never just present or absent, but always occurs as a happening (*Geschehnis*). While the temporal character of all these elements seems clear, the nature of their temporality requires further elaboration.

Truth in the Artwork leads to the Decision

The question now becomes: how does the truth come to pass in the artwork? We have seen that it is not by corresponding to a "true" state of affairs or by "truly" describing an object (in the sense of *Richtigkeit*) that truth sets to work in the artwork, but by clearing a space for beings through the instigation of the conflict between world and earth.

> In dem Streit wird die Einheit von Welt und Erde erstritten. Indem eine Welt sich öffnet, stellt sie einem geschichtlichen Menschentum Sieg und Niederlage, Segen und Fluch, Herrschaft und Knechtschaft zur Entscheidung. Die aufgehende Welt bringt das noch Unentschiedene und Maßlose zum Vorschein und eröffnet so die verborgene Notwendigkeit von Maß und Entschiedenheit. [UK,49]

Heidegger paradoxically insists that unity is only the result of conflict, i.e., disunity; it is *erstritten* in the way a victory can only be achieved through a battle. The conflict here is not just any conflict though, but rather the conflict between world and earth, i.e., between the possibility of there being anything at all and the one "thing" that comes forward as something. It is only within a particular world that concepts like victory and defeat, blessing and curse, mastery and slavery, take on meaning. These are not just arbitrary dichotomies which one can choose to recognize or not. They are "given" along with the world that the work has opened up. Through the conflict, a unity or mutual belonging-together of world and earth is maintained, in which what is still up for decision, whatever seems unruly and chaotic, can be seen to arise from previously tacit decisions.

The key term in this passage is the "decision" (*Entscheidung*). It is that which gives a world its form and meaning, its legacy and its possibilities. The world is found unfinished, it still demands decision and measure from those who inhabit it. When a world is opened up, decisions get made. Even though the decision is equiprimordial with the "discovery" of the world, Heidegger implies that the (or some) decisions are still open. He says "victory" and "defeat," etc. are "put up for decision": *"Indem eine Welt sich öffnet,* stellt *sie Sieg und Niederlage ...* zur Entscheidung." He then subtly modulates the term *Entscheidung* into *das noch Unentschiedene* ("the—as yet—undecided"), that which is still outstanding. This maneuver makes possible a double reading of *Entschiedenheit* as "decisiveness" or "decidedness." That is, *Entschiedenheit* taken by itself would normally be read as a psychological state of mind: e.g., *"Er schrieb mit Entschiedenheit an seinen Vorgesetzten."* By its proximity to the nominalized past participle, *das Unentschiedene*, however, it takes on the flavor of "the state of having decided." As a result, it can be used to refer to the consequence of the recognition of a need for decision.

Heidegger sees a world opening up by forcing a decision (or a set of decisions) on a particular "historical mankind" (*geschichtliches Menschentum*). This formulation explicitly includes all human beings; no perspective on "peoples," "races," or other subgroups is given. Rather, what one belongs to is not this or that ethnic or national group but to a particular historical moment, age, or tradition. All of those partaking of the "history" of their group during the time in which the decision is open to be made make up the "historical humanity" which is the operative entity. Heidegger thus seems to confront a *generation* with its decisions concerning "victory" and "defeat," "blessing" and "curse," "mastery" and "subjection." History is not a simple drama or merely the result of mechanical forces, but a way of answering certain basic questions like: Who are truly my contemporaries? Where do we seek victory or fear defeat? What counts for us as a blessing, and how do we distinguish it from a curse? Because these are indeed *basic* questions, they *always get decided*, one way or another. Choice exists as among alternative paths, but not as to whether or not to choose. At the same time, it is impossible to say "who" chooses a particular path. Since those affected by the decision are defined through belonging to a particular historical time, we may say that *the decision then is the distilled essence of time.* It is time understood as the basis of history.

It is crucial for understanding the importance of the decision (*Entscheidung*) for Heidegger's thought to avoid conflating it with the more well-known "resolutness" (*Entschlossenheit*) discussed in *Being and Time*. Heidegger himself goes out of his way to distinguish the two concepts.

> Die in "Sein und Zeit" gedachte Entschlossenheit ist nicht die decidierte Aktion eines Subjekts, sondern die Eröffnung des Daseins aus der Befangenheit im Seienden zur Offenheit des Seins. [UK,53]

This passage clearly seeks to distinguish *Entschlossenheit* from any subjectivistic misreading that might arise from the commonly-used verb *sich entschließen* (to decide or resolve). Even the emphatic contrast between the Latin terminology on the one hand (*"die decidierte Aktion eines Subjekts"*), and the etymologically transparent (to a German-speaking audience) Germanic roots on the other, underscores this difference. Heidegger is not engaging in a discourse within the Western philosophical tradition (represented by the metaphysically burdened latinate terms), but attempting to reach an understanding of a problem. He contrasts the traditional conception of *Entschlossenheit* with the understanding which frees Dasein from subjectivity and thus "unlocks" (*ent-schließt*) Being for Dasein in its potentialities. *Ent-*

schlossenheit opens up the world for Dasein and thus discloses Dasein's ownmost possibilities.

The decision referred to by *die Entscheidung*, however, rather than opening man to Being, *cuts him off* from possibilities and "determines" his temporality. Man may indeed have an access to Being, but he is not thereby freed from time. When the decision has been made, no matter how or by whom, a world opens up and comes to take on a sharpness of outline ("*die unnachläßliche Schärfe der Beirrung*") that distinguishes the "real" from the "unreal," the "true" from the "false." In this way the decision of an historical humanity, although arising from its past, cuts it off from that past and starts the essential conflict of world and earth anew. The decision with which the artwork confronts those who partake of it "transports" them (to use Hofstadter's term[11]) out of their familiar surroundings. Gerald Bruns emphasizes this dynamic and even violent quality inherent in great works of art.

> The occurrence of truth always means estrangement from what is merely present and familiar. . . . It is rather as if "the immediate circle of beings" in which we feel ourselves so comfortably disposed were not a world after all but a closure which it then becomes the task of art to break open and perhaps even to bring crashing down.[12]

On the other hand, the truth of the artwork also calls forth decisions from which the future history of a people can emerge. The "state-founding act," to use one of Heidegger's examples, obliterates what went before by turning towards new possibilities which it has just opened up. The work, while destroying old stifling barriers and conventions, never delivers us out of *all* worlds. Its freedom is never anarchy.

The decision, however, still calls for someone to make it. Who else could make it, if not those who choose to stand (or find themselves standing) within its world, the world of the artwork? Heidegger calls them the "preservers" (*die Bew*ahr*enden*) of the work, thereby connecting them with the "truth" (*die* Wahr*heit*) of the work.

> Bewahrung des Werkes heißt: Innestehen in der im Werk geschehenden Offenheit des Seienden. Die Inständigkeit der Bewahrung aber ist ein

> Wissen. Wissen besteht jedoch nicht im bloßen Kennen und Vorstellen von etwas. Wer wahrhaft das Seiende weiß, weiß, was er inmitten des Seienden will. [UK,53]

The preservers of the work, those who have stepped into its world, already have a kind of knowledge of that world, but it is not the connoisseurship of an expert or the self-indulgent phantasies of a dilettante. Anyone who has been truly affected by a work of art *knows* something, even if it cannot always be put into words satisfactorily. Further, that knowledge has to do with a kind of willing which, as we have seen above, is not to be confused with *"die decidierte Aktion eines Subjekts."*

> Das Wissen, das ein Wollen, und das Wollen, das ein Wissen bleibt, ist das ekstatische Sicheinlassen des existierenden Menschen in die Unverborgenheit des Seins. [UK,53]

> That knowing which is also a willing and that willing which stays a knowing is the ekstatic letting-oneself-into the unconcealedness of Being by existing man.

This is by no means an unambiguous statement. To begin with, the term *"Mensch"* could refer to a single human being or to the *gens humana*. Further, what I have translated as "letting-oneself-into" (*Sicheinlassen*) could equally well be translated as "involvement in" (the unconcealedness of Being), although I do not see how Hofstadter finds a "compliance" in this sentence, since no demand is being made here.[13] Heidegger seems to be saying that the kind of knowing that goes on when we come under the sway of the work of art implies an antecedent engagement with the work, so that it is able to matter to us. Such is the essential hermeneutic position required by the work. No one can be said to have truly "objective" knowledge, in which the thing known is remote from and uninvolved with the knower.[14]

Nor is "willing," as Heidegger understands it, an imposition by an isolated, independent subject on something external to it. It is not aggressive or outward-reaching. Rather, willing requires that we open ourselves up to and expose (*aussetzen*) ourselves to the work. Willing in this context is actually closer to receptivity or even vulnerability than to self-assertion.

> Wollen ist die nüchterne Ent-schlossenheit [sic] des existierenden Übersichhinausgehens, das sich der Offenheit des Seienden als der ins Werk gesetzten aussetzt. [UK,54]
>
> Willing is the sober [i.e., "thoughtful," "perspicacious"] resoluteness of that being which in existing transcends itself, [and] which is exposed to the openness of beings in the form which that opennesss has been given in the work.

Willing, by a contradictory move, is both self-possessed (*nüchtern*) and transported out of itself (ek-static); it is not intrusive but instead receptive to the kind of way beings come forward (appear) in an artwork. At the same time, willing also belongs to the maneuver by which entrance into the clearing of the artwork is attained, as the previous passage showed. It is through this opening up to the work (in its widest sense) that people recognize themselves as belonging together and thus as historical. Through the work of art they reach a decision which enables them to "step into" their own history.

> Wissen als Gesehen-haben ist ein Entschiedensein; ist Innestehen in dem Streit, den das Werk in den Riß gefügt hat. [UK,54]

Entschlossenheit, by opening up to what is happening in the work, enables a "willing," i.e., things in the work come to have meaning and value, we can no longer be indifferent to them. "Knowing" has "seen" something and thus has already cut up the whole in some way. It is thus *"ein Entschiedensein"*: [15] *After* I have let the work matter to me—and thus "speak" to me—, *after* I have recognized it as something important, I "know" something, even if that is merely that it is "good" or "bad." A decision has thus already been made, though it is not "I" who has made it; rather, it was made *for* me by the resoluteness which allowed me to open to the work in the first place. But knowing does not just know empty temporal categories. It also discloses what is/was/will be important. Thus Heidegger insists that *"Whoever truly* knows *an entity, knows what he* wants *from it."* [UK,53, emphasis added] By disclosing what "counts" or "matters," such knowing opens a space for truth to "happen" (*geschehen*) and thereby reveals history (*Geschichte*) as a "happening."

> Die Bewahrung des Werkes vereinzelt die Menschen nicht auf ihre Erlebnisse, sondern rückt sie ein in die Zugehörigkeit zu der im Werk *geschehenden* Wahrheit und gründet so das Für- und Miteinandersein als das *geschichtliche* Ausstehen des Da-seins aus dem Bezug zur Unverborgenheit.
>
> [UK, 54; italics added]

This passage emphasizes that we do not come upon an artwork as isolated subjects who receive some experience from it. Those who truly hear or respond to the artwork are actually its "preservers," who, by stepping into its "working truth" (*"der im Werk* geschehenden *Wahrheit"*), step into or open up a historical relationship to what matters (their "truth"). Such a truth, it must be emphasized, only appears when the decision has been taken (an *Entschiedensein*).

Poetry and Truth

"Der Ursprung des Kunstwerkes" began with the assertion that artwork and artist have their origin in art. At the time, the term "art" had to remain somewhat vague, but after the discussion of truth and the decision in the artwork, Heidegger returns to it and offers the following quasi-definition:

> Kunst ist das Feststellen der sich einrichtenden Wahrheit in die Gestalt. Das geschieht im Schaffen als dem Hervor-bringen der Unverborgenheit des Seienden. Ins-Werk-Setzen heißt aber zugleich: in Gang- und ins Geschehen-Bringen des Werkseins. Das geschieht als Bewahrung. Also ist die Kunst: die schaffende Bewahrung der Wahrheit im Werk. *Dann ist die Kunst ein Werden und Geschehen der Wahrheit.*
>
> [UK,57; italics in original]

What is important in the present context about this definition is its introduction of temporality into art. Art is not an immutable form or image (*eidos*), but a *"becoming and happening of the truth."* The work lets beings approach us as themselves without either being emptied into mere "things" (*Vorhandenes*) or disappearing into a reliable functionality as equipment (*Zuhandenes*). In everyday life we hardly notice beings as such; the nature of equipment ensures that the better it functions, the more invisible it becomes. Its serviceability brings it to rest in the everyday and brings us to rest—i.e., removes us from time—in the everyday, too. Art, however, removes things from their invisible serviceability and throws or projects (*entwirft*) them into the clearing of the work. The "projecting" of truth occurs as "poetizing" (*Dichtung*).

> Wahrheit als die Lichtung und Verbergung des Seienden geschieht, indem sie gedichtet wird. *Alle Kunst* ist als Geschehenlassen der Ankunft der Wahrheit des Seienden als eines solchen *im Wesen Dichtung*.
>
> [UK,58; italics in original]

In this too-famous passage, Heidegger seems to assert a priority of poetry over other forms of art. He does not immediately clarify what he means by "poetry" (*Dichtung*) and in fact frequently uses the word as verb: *"gedichtet [wird]," "[aus dem] dichtenden Wesen der Kunst,"* etc. While as verb, *dichten* could be translated as "poetizing," i.e., "making poetry," the word is also related to the Latin *dictare*, "to say for transcription, dictate."[16] Art is thus a *saying*, one which aims at permanence (hence the transcription). The key to art as *dichtend* is its ability to transform the everyday into something else.

> Aus dem dichtenden Wesen der Kunst geschieht es, daß sie inmitten des Seienden eine offene Stelle aufschlägt, in deren Offenheit *alles anders ist* als sonst. Kraft des ins Werk gesetzten Entwurfes . . . wird durch das Werk *alles Gewöhnliche und Bisherige* zum Unseienden. [UK,58, italics added]

The artwork now turns out to have something in common with Heidegger's own method. It was mentioned before that his hermeneutic begins with the everyday, the ordinary, with simplistic clichés, with what occurs *"proximally and for the most part"*[17] (*zunächst und zumeist*). *"Der Ursprung des Kunstwerkes"* began in this way with naïve questions about the relationship of artwork and artist and with a conception of the artwork as "thing." Viewed as "things," a painting is quite different from a temple or a poem. They differ in form and content. But that is taking them as "things" without first understanding what a "thing" is. It is rather through the artwork that the thing becomes apparent in itself. Through the artwork *"everything is different from before."*

Although he at first denies any kind of priority to *Dichtung* in its narrow sense as verses ("*Poesie*"), Heidegger soon concedes it particular importance as a "linguistic artwork." Language is much more than a means to communication or a series of sounds:

> Sie befördert das Offenbare und Verdeckte als so Gemeintes nicht nur erst in Wörtern und Sätzen weiter, sondern die Sprache bringt das Seiende als ein Seiendes allererst ins Offene. [UK,59]

According to Heidegger, only in language can beings be accessible to us *as* beings, as what they "truly" are (*in Wahrheit*). The ability of language to provide a version of what it means to be he calls *entwerfen*, casting forth, in the sense of "pro-jecting." *Dichtung* thus brings to light the essential trait of language, which is irreducibly historical.

> Die Dichtung ist die Sage der Unverborgenheit des Seienden. Die jeweilige Sprache ist das Geschehnis jenes Sagens, in dem geschichtlich einem Volk seine Welt aufgeht und die Erde als die Verschlossene aufbewahrt wird. [UK,60]

The emphasis in the second sentence above, I believe, should fall on the word *"geschichtlich"* which is used adverbially instead of modifying *Volk*, as would be expected. Heidegger does not say here that a particular people (such as the German people) stands at the center of history, but that for each people a language occurs *historically* and through its occurrence (*Geschehnis*) reveals its world, while simultaneously preserving the hiddenness of its earth. We must also not overlook the alternation of *die Sage* with *das Sagen* when Heidegger talks of language. Without making an issue of it, he hints at the narrative nature of language and underlines its role in making history. As opposed to mere "saying" (*das Sagen*), a people's "saga" (*die Sage*) is always central in its self-understanding as a people. Also, it is plain to see how near to each other poetry and language are. In poetry as artwork, the truth of language comes forth as nowhere else, even though language itself is at bottom *Dichtung*:

> Die Sprache ist nicht deshalb Dichtung, weil sie die Urpoesie ist, sondern die Poesie ereignet sich in der Sprache, weil diese das ursprüngliche Wesen der Dichtung verwahrt. Bauen und Bilden dagegen geschehen immer schon und immer nur im Offenen der Sage und des Nennens. [UK,60]

What is a temple, for instance? For birds it is a place to build nests, for dogs perhaps a place to relieve themselves, for rodents a hiding place. Without the "clearing" made possible through language, the temple recedes into "mute stones." It loses its status—or rather, activity—as artwork and as cultural locus. Poetry derives its special power from its closeness to the fundamental nature of language in human existence. Poetry (as *Dichtung*) sets out the parameters within which "building" and "images" (and "words") can have meaning.

One might be tempted to ask why poetry is given such a privilege in relation to language. Are not philosophy, mathematics, history, especially "foundational?" Heidegger would agree that they too can serve as works, but would add that they are derivative. Only poetry preserves the "original sense" of *Dichtung*. Still, *Dichtung* is not an "urpoetry": Heidegger does not want to romanticize the earliest examples of linguistic artworks as the most important. *Dichtung* does not refer solely to verses. Its value derives from its ability to found (*stiften*) truth. Heidegger distinguishes in *Dichtung* a threefold *Stiftung* of truth: as giving or granting, as founding, and as beginning.

> Das Wesen der Dichtung aber ist die Stiftung der Wahrheit. Das Stiften verstehen wir hier in einem dreifachen Sinne: Stiften als Schenken, Stiften als Gründen und Stiften als Anfangen. Stiftung ist aber nur in der Bewahrung wirklich. So entspricht jeder Weise des Stiftens eine solche des Bewahrens. [UK,61]

We have already seen that the artwork sets the truth to work. Here Heidegger expands on that statement by assigning foundational granting to *Dichtung*. Once again Heidegger seeks to combat the tendency to view art as form and content. While the artwork each time comes from "somewhere," i.e., from materials in a broad sense, as well as from the artist, it can in no way be derived from them or from anything else that went before. The stones do not presage the temple. To know, for instance, that Leonardo studied with Verrocchio could never be construed as allowing us to predict the kind or quality of Leonardo's paintings or even that there would be paintings. They are always freely "given" (*geschenkt*):

> Die im Werk sich eröffnende Wahrheit ist aus dem Bisherigen nie zu belegen und abzuleiten .... Die Stiftung ist ein Überfluß, eine Schenkung.
> [UK,61]

In a sense, it will always remain a mystery for us, Heidegger says, whence artworks come; we can only call them a "gift" (*eine Schenkung*).

We must not think, however, that the artwork comes from a "gift" in the sense of the artist's "gift" (i.e., his talent). Whatever is given comes from the basis (*Grund*) of his and our very existence. Therefore, since *Dichtung* brings to light what was hidden in our world, what Heidegger earlier called "earth," its founding also has the character of a "grounding" (*Begründen*). Here Heidegger's usage is closer to the English sense of "grounding" meaning "having one's feet on the ground." *Dichtung* grants authority to the artwork by drawing its truth from the "ground" of the world.

Besides being a granting and a founding, *Dichtung* as *Stiften* also gives the artwork its immediacy (*das Unvermittelte*), that which makes it unique and a beginning (*Anfang*). Its immediacy comes from *Stiften* as *Anstiften*, from "starting" as a kind of "starting up" or instigating (e.g., a fight): *"Kunst als Dichtung ist Stiftung in dem dritten Sinne der Anstiftung des Streites der Wahrheit, ist Stiftung als Anfang."* [UK,62] Heidegger again reminds us of the discussion of truth in the artwork. Truth was the bringing into the open of the conflict of world and earth. Through *Dichtung*, art is able to start up the conflict again, as a new beginning. When art is no longer able to do that, it is dead: its world has vanished and its earth withdrawn into the everyday.

As such a beginning, art is historical in the sense of "historic." History is not just a chronological ordering of facts, but an event, as we say of something particularly important that it was a "real event," or "historic," or "epoch-making."

> Immer wenn Kunst geschieht, d.h. wenn ein Anfang ist, kommt in die Geschichte ein Stoß, fängt Geschichte erst oder wieder an. Geschichte meint hier nicht die Abfolge irgendwelcher und sei es noch so wichtiger Begebenheiten in der Zeit. Geschichte ist die Entrückung eines Volkes in sein Aufgegebenes als Einrückung in sein Mitgegebenes. [UK,63]

Thus art does have a relationship to history, but not as usually understood by art historians. Rather than seeing historical determinants of art, Heidegger sees art making history itself possible. Christianity did not bring forth the cathedral, nor did the Industrial Revolution bring forth Social Realism in painting. The cathedral gave medieval Christianity a place to come forward as fully as possible, to bring the conflict of world and earth into a new "light." The paintings of Courbet bring the nature of science and industry into clearer and more meaningful focus than science and industry themselves. As Robert Bernasconi remarks,

> To speak here of "the historical nature of art" means something very different from referring art to history as if in an attempt to explain it. Art itself is shown to be one of the ways in which history takes place. And history is neither the work of man, nor still less a straightforward chronology. History is understood as a rupture. It takes place in rare epoch-making events that determine what is and what is not....[19]

History is the volcano which sleeps for ages and then suddenly "erupts." Great art, like history, is seldom. When it occurs, whether as culmination or as revolution, it alters the landscape to such an extent that a return to the old ways and times is impossible. Thus are epochs made.

The last sentence in the above quotation from Heidegger needs to be emphasized. History brings a people before that which has been given to it to accomplish (*sein Aufgegebenes*), the "self-evident" tasks to be undertaken, problems to be solved, danger to be avoided. No people can choose what these will be; all are "carried off" (*entrückt*) into them. This carrying-off[20] occurs as a carrying-back to something that had been "there" all along, as something which had been "given" to the people from the very beginning. Thus, history occurs as if one were transported back to one's roots, to that which is most important to one's own particular people. In that way, history is always conservative, it always brings one back to the forgotten founding.

In the last few paragraphs of the essay, Heidegger summarizes briefly what has been discovered about art. He then goes on to bring art and his discussion of art back to its origin (*Ursprung*). What is the origin of the artwork? We have by now learned to hear better what the word "*Ursprung*" says:

> Die Kunst läßt die Wahrheit entspringen. Die Kunst erspringt als stiftende Bewahrung die Wahrheit des Seienden im Werk. Etwas erspringen, im stiftenden Sprung aus der Wesensherkunft ins Sein bringen, das meint das Wort Ursprung. [UK,64]

This nearly untranslatable passage completes the circle on which we began. Art brings truth out in a leap or rush (*Sprung*). The rush is the immediacy of the artwork, which seems to be offered to us suddenly and inexplicably from "nowhere." In this onrush from nowhere, art brings out the truth of what is (*die Wahrheit des Seienden*). The word "origin" means, according to Heidegger, *etwas erspringen*, which could be translated as: to "jump start" something, *"to bring something suddenly out of its hiddenness in the tradition into the light of Being."* The origin of art comes from its ability to originate. Yet, it does not strictly originate *ex nihilo*: art brings together into conflictual harmony Being and the tradition. It brings the tradition to life and keeps it alive. Art is the origin of the artwork,

> ...weil die Kunst in ihrem Wesen ein Ursprung ist: eine ausgezeichnete Weise wie die Wahrheit seiend, d.h. geschichtlich wird. [UK,64]
>
> ...because art is essentially an origin: an outstanding way truth comes to be, i.e. becomes historical.

An origin is a place where action begins, where what was hidden or non-existent comes to be (*seiend wird*). When truth comes to be it enters history, it "becomes historical" (*geschichtlich wird*). After that point, it can be approached by entering the world of the artwork.

We have not yet reached the conclusion of the essay. The purpose of these lectures was to inquire after the origin of the artwork. The goal was, however, not just to "find" the origin, in the sense of bringing it out of its invisibility, but to be better able to inquire further

> ...ob die Kunst in unserem geschichtlichen Dasein ein Ursprung ist oder nicht, ob und unter welchen Bedingungen sie es sein kann und sein muß.
>
> [UK,64]

Having found the origin of the artwork does not answer the question but rather confronts us, Heidegger's hearers and readers, with our own question. Art asks *us* if it is still "originary" in our history. Is it still capable of being an origin? Or has it been reduced to a theoretical knowledge, one which knows "how" but not "why," or has it become a self-satisfied connoisseurship? This should not be construed as a challenge but as a "decisive" question, a question which calls for a decision.

> Sind wir in unserem Dasein geschichtlich am Ursprung? Wissen wir, d.h. achten wir das Wesen des Ursprunges? Oder berufen wir uns in unserem Verhalten zur Kunst nur noch auf gebildete Kenntnisse des Vergangenen?
>
> Für dieses Entweder-Oder und seine Entscheidung gibt es ein untrügliches Zeichen. Hölderlin, der Dichter, dessen Werk zu bestehen den Deutschen noch bevorsteht, hat es genannt, indem er sagt:
>
> "Schwer verläßt
> Was nahe dem Ursprung wohnet, den Ort."
>
> *'Die Wanderung,'* Bd.IV (Hellingrath) S.167
> [UK,64]

"Der Ursprung des Kunstwerkes" thus closes[21] with a turning to the audience and to Hölderlin. The audience, Heidegger included, faces an "either/or" about history and about art, but it seems unlikely that

this is a true "issue," one still up for decision; surely it is an implicit *demand* that art be an origin again. Heidegger here confronts us with our own choices, the "prejudices" (*Vorurteile*) of our own history as we read it. Modern man has indeed wandered far from his origins—often he glories in the knowledge that he has come so far. But man's origin is also his center, the source of stability and meaning. Art is one of the "outstanding ways" of returning to this source and replenishing or founding again human life.

By the end of the lecture we have arrived at some verses of Hölderlin. His two lines serve as a crucible for the decision demanded just before. They sound mysterious, hermetic, they seem to describe some gravitational pull to the origin. Whether this origin is Heidegger's *Ursprung* or not, we cannot say. Only when we read *"Hölderlin und das Wesen der Dichtung"* will we find out how important Hölderlin is for Heidegger's concept of the origin of the artwork.

Fazit

Has Heidegger done what he set out to do? Has he destroyed the viability of the form/content dichotomy in the discussion of artworks? He has certainly shown that there is another way of talking about artworks and that it is inextricably bound up with the basic issues of human existence and history. But has it displaced the traditional, "aesthetic," approach to art?

What he has established, I think, is the importance of choosing the right approach to understanding an artwork. By its claim to self-sufficiency the artwork demands to be taken on its own terms. *"Der Ursprung des Kunstwerkes"* represents an attempt to lay out in general what those conditions of approach could look like.

Since temporality is the *Leitwort* of our own study, we have paid particular attention to Heidegger's method of approaching the artwork, especially its hermeneutic aspects. The work of art is not accessible to our grasp—it is like a trick candle that cannot be blown out. Each time we approach it, it recedes. At the same time, the artwork is so immediate that it cannot be avoided. It thrusts us before the question of our relationship to our own time. Is it an "original" relationship or only "scholarly trivia" (*gebildete Kenntnisse*)?

The question remains, however, as to how this new understanding of the artwork actually functions when confronted with a major piece of poetry. Heidegger has stayed rather distant from *linguistic* artworks here, as though deliberately seeking to avoid confusion when introducing new terms and a new way of looking at art. It is, after all, less misleading to speak of the conflict between "earth" and "world" when the artwork comes from the plastic arts, than in a text, where the same words could be used in their ordinary sense. And yet, given the prioritization of *Dichtung* over against other artworks, the natural next step would lead to a discussion of a poetic text.

Heidegger himself, however, tackled the general problem of the status of *Dichtung* once again before turning to extended *Erläuterungen* of individual poems by Hölderlin. In his lecture, *"Hölderlin und das Wesen der Dichtung,"* given in Rome in 1936, shortly after the lecture series which became *"Der Ursprung des Kunstwerkes,"* Heidegger expanded on the importance of poetry, not only for understanding the nature of the artwork, but for understanding the kind of relationship man has to his own Being. Choosing words of Hölderlin, he set out to remove any prejudice in favor of the "classics," or "great works," which might have remained from the examples used in the former lectures. The search for the "origin" of the artwork has led him to the "essence" of poetry (*das Wesen der Dichtung*). Heidegger was not making incidental remarks on a topic of literary or aesthetic interest. He set as his goal the fundamental status of the whole field, its importance to historical humanity.

If we glance back at this point to the problems encountered in our initial reading of *"Wie wenn am Feiertage...,"* it may seem that the poem has not advanced, but receded from our view. Where before we seemed to have a text before us, solid and reliable, even if difficult to read, now we have lost the very key which was to open it for us: temporality. Indeed, the text itself seems much more problematic. It is no longer susceptible to analysis based on linguistics or stylistics, history or biography. The terms used in our analysis have been shown to be empty, based on a mistaken understanding of the nature (i.e., origin) of the artwork.

Heidegger has set the artwork onto a basis which we can indeed call temporal, in fact historical, but only when these words are radically reinterpreted. A poem is able to speak in time and about time, but it cannot choose its own time. It rather springs from its time as from its "earth," and gives form and meaning to that time through its *"stiftenden*

Sprung." Our first mistake was to try to take the poem apart (analysis), which would effectively return it to its elements and precisely *overlook* its "Word." Instead of taking it apart into its elements, Heidegger seeks to hear its originary speaking. Thus, while the text as "words" may have receded, the text as *Dichtung* now begins to approach. It remains to be asked, though, "What is the nature of *Dichtung*?"

Chapter Four

"Hölderlin und das Wesen der Dichtung": POETRY AND THE DECISION

THE READING OF *"Wie wenn am Feiertage..."* advanced in Chapter Two followed the temporality of the poem into an aporia in which the ultimate standpoint of the poet was found to be undecidable. Is this a sign of the poem's "failure," that the poet did not have command over his resources? It does not seem so. The other side of the poem's difficulty and inaccessibility—its attractiveness—remains untouched by our attempts to reach it.

We sought the temporality of *"Wie wenn am Feiertage..."* as though it were something which could be determined and settled, which could be made precise and stable, once and for all. Our focus on the linguistic and structural markers of temporality in the text subscribed to a science-like methodology which sought general and repeatable results. Such an approach was flawed from the outset because it neglected the temporal nature of the very thing sought. Temporality must not be sought as though it were similar to historical or literary references, for example. Careful study is able to reveal historical references—say, to the French Revolution—or mythological references—e.g., to Semele, Zeus, and Dionysos—or literary influences—to Pindar, Klopstock, Schiller, etc.—all of which, once found, remain stable: once "discovered," they do not change. Temporality, on the other hand, involves change, process, direction, beginning, ending. While these can be encoded in the structure and content of the words in a text, they themselves escape such a search, which is similar to an archeologist's searching for a vanished culture solely through an investigation of surviving artifacts. Temporality is precisely what cannot be "found" in a text; it can only be *inferred* from the results of reading. Perhaps the problem was with our approach. It seems appropriate to rethink our method of reading.

At the end of *"Der Ursprung des Kunstwerkes"* Heidegger quoted two lines from Hölderlin's poem *"Die Wanderung."* He claimed for these lines, as for Hölderlin himself, a privileged place in the historical situation of the German people. He made quite clear that Hölderlin's words bear significantly on the outcome of the decision about the "place" occupied by the German people in his day (1936). *"Sind wir in unserem Dasein geschichtlich am Ursprung?"* Hölderlin has "named" the "unmistakable sign" about that question. Yet his verses are far from unambiguous. They seem to say that one who dwells near the "origin" only leaves it with difficulty. That, however, says two things: it is hard to leave the origin, but one *must* anyway. The undeniable power of the lines seems to come partly from their dark, enigmatic sense, and partly from the corresponding awkwardness of their rhythm, with its sprung word order and elided subject. If these verses of Hölderlin do indeed give a "sign," it remains yet to be deciphered. Are they perhaps, in their simplicity and obscurity, the sign to which Heidegger is referring? That is, they seem to point—this much is clear—to Hölderlin as a possible pathbreaker into the nature of poetry.

In *"Hölderlin und das Wesen der Dichtung,"* the most programmatic of the essays in the *Erläuterungen*, Heidegger addresses the question of poetry as a whole and of Hölderlin's place within poetry. It turns out that poetry has a central role to play in human existence and that Hölderlin has a key role to play in talking about poetry. Thus, Heidegger seeks the truth of Hölderlin's poetry, which brings him to the truth of poetry itself, which grows out of an insight into the nature of truth.

Throughout, it is his *way* of reading Hölderlin that distinguishes Heidegger's approach. He does not seek an atemporal concept to stand for Hölderlin's texts. By responding to them in his/their own way, Heidegger allows their temporality to guide and direct his own reading. He thus opens a dialogue (*Gespräch*) with the texts, and we shall see that the dialogue as part of the hermeneutic process takes an important place in Heidegger's thinking about poetry.

The essence or center of Heidegger's essay is indeed the *Wesen der Dichtung*, as the title implies. However, although it is the center, this "essence" only emerges fully at the text's end. To understand how the "center" can be found at the "end," attention must be paid to the temporal structure of the text, found through the process of reading. The text is not some kind of conveyor belt, delivering a finished product at the "end" of the line; nor is it a joke—we get no "punch line." On the other hand, it is also not just a catalogue of highlights. "The essence of poetry," in short, belongs to the text as its *telos* or end. It will be necessary to traverse a certain path before the "end" can be reached. The path is marked out in advance by the five "leading words." In what way do they lead us?

The first two leading words, though stemming from one and the same author, lead immediately into a contradiction: poetry is both "the most *innocent* of occupations," as well as a manifestation of language, the "most *dangerous* of goods." Although the contradiction is recognized, no attempt is made to resolve it. Rather, Heidegger proceeds to ask three *Vorfragen* which in turn lead into a more fundamental reading of "innocence," "danger," "language," and "poetry." By understanding these words, Heidegger is able to return to the initial contradiction and to the radical assertion with which the essay began (see below, p. 104). The "end" of the essay is a better understanding of its beginning (*Ursprung*).

Such a method is not dialectic, since it does not attempt to harmonize the contradictions; they are not merely "tools" to be used towards a predetermined goal. Precisely this conflict in the words of the poet is what we seek to understand. Nor does the reader look down on the contradictions from a "higher" understanding. Nor is this a linear journey "through" the text, brushing aside any contradictions as inessential. Whereas dialectic leads ultimately to cessation or rest in the Absolute, Heidegger's hermeneutic ensures that the process of reading can never rest in self-satisfaction. I believe Heidegger has seen that reading with excessive haste leads to viewing words as only hindrances to a goal (truth, knowledge). He teaches reading thoughtfully, in which that which is to be thought does not recede into a larger form or purpose but remains with its thinking without regard to the "time" required. Such thinking allows that which is to be thought to set the time, rather than importing an extraneous, "objective," time. This in turn enables the "essence of poetry" to step forward more clearly. As Robert Mugerauer has observed in another context,

> The fruitful way to get at the whole work involves our overcoming our urge to get quick results by finding an assertion to hold on to, carry off, and verify. Thus, meditative thinking, the thinking perhaps appropriate to the nature of thought . . . would involve an active waiting . . . .[1]

Our own "method" starts with a look at the structure or process of Heidegger's thought. Structurally, the essay begins with a question (*"Warum ist ... Hölderlins Werk gewählt?"*), and then questions it in turn (*"Weshalb nicht Homer oder Sophokles...?"*). As a response to these questions, he returns to the word *Wesen* in the title and brings it forward by distinguishing the "essential" in essence from the merely "general," which is always "inessential." The recursive, self-reflective course of Heidegger's thought already shows itself. Each step of the search for the *Wesen der Dichtung* leads back to a previous word. Although it may appear that nothing gets truly "demonstrated," that each step of the exposition merely places into question the previous one, in fact understanding grows and intensifies through the process. Difficulty only arises when the search is seen as linear and the goal as atemporal. As Heidegger stresses, however, the "essence of poetry" is not to be found in a word.

"Hölderlin und das Wesen der Dichtung" is signficant for this study as a model of reading which does not ignore the effects of temporality, either in the text itself or in the method of interpretation. Heidegger claims to have discovered in Hölderlin's writings the "essence" of poetry. Can we too, as readers, discover this essence by reading Heidegger's text? And can we then turn to Hölderlin's poetic text and discover its essence anew? To answer that question, we must once again begin with the naïve, the superficial, with what is most obvious—and use that as a beginning. I will look at Heidegger's *process of reading* in his encounter with Hölderlin. I thus diverge from the majority of Heidegger's interpreters, who often are so involved in explaining his terms that they overlook the *way* in which those terms are used, the hermeneutic ground of his language. Focusing on the *process* of interpretation, I hope to remain true to its temporal nature.

Judged from a temporal point of view, each of the five *Leitworte* envisages the importance of time for poetry and for man. Poetry brings forth the bare possibility of a time and simultaneously makes it accessible to man. Through the poetic word, humans are able to come together in dialogue and the essential decisions about man and his place in the world are made. Each leading word, although deliberately stripped of its original context, does indeed lead to some undeniably important aspect of Hölderlin's thought. Heidegger does not *want* us to "understand" these words, to find a comfortable niche for them in our historical treasure-house; they are too easily overlooked that way. To say that the five words do not add up to a coherent whole or systematic unity is merely to state the obvious. On the other hand, they clearly have not been chosen arbitrarily, but actually show themselves highly "thoughtworthy."

Heidegger continues in the same fashion throughout the essay, advancing a statement, questioning it until the contradictions (or just peculiarities) inherent in the relationship of question and answer give rise to a different question and thus to a different possible answer. In this case, the contradictions between the nature of poetry as ordinarily conceived and as it appears at special times (e.g., at a wedding or funeral, before going into battle, when one is in love) lead to another look at poetry—from the ground up, so to speak. We discover that it is through language that poetry attains its innocence, as well as its effectiveness.

A Word about Method

On such a peculiar search, it is important to be clear about the method employed. Heidegger's hermeneutic initally approaches a text as something which is always already understood in some sense. Such an understanding is an "everyday" one, "anyone" could have it. It serves to keep the text at a distance as something used or contemplated. Everyday understanding of the text, while incomplete of itself, suffices as a beginning to lead to a more fundamental questioning of the text, if it is not hastily forgotten in looking for something "new." How can everyday understanding become "essential" understanding (*wesentliches Verstehen*)? Only by allowing that which is to be understood to matter to us, by recognizing the reality of the words in the text and allowing them to speak to us. Such is the nature of the dialogue, as Heidegger describes it here (see below, pp. 116 –120).

As Heidegger has chosen five "leading words" to guide his reading, we too will choose some words to lead ours. The first word is "searching" (*suchen*), a temporally explicit way of encountering a text. Heidegger frequently refers to his activity as a search or path.[2] E.g., "*Doch eben dieses Wesentliche des Wesens* suchen *wir...*"[HWD,34] As a search, we will not proceed linearly, not being guided by external absolute standards, but recursively, letting the text itself set the direction. A search is different from a journey. A journey has a fixed goal and can therefore have a clear, linear itinerary. A search begins with an absence and proceeds circumspectively. Each successive clue or signpost points not only forward but backward as well. It must be brought into relation with the starting-point to be useful in reaching the goal. Thus, our search must relate each step to the previous one as well as to the total project.

The second word is "decision" (*die Entscheidung*): Heidegger's essay begins by asking about the choice of Hölderlin. The answer provides the clue to our reading of Heidegger and Hölderlin.

> Doch eben dieses Wesentliche des Wesens suchen wir, jenes, was *zur Entscheidung zwingt*, ob und wie wir die Dichtung künftig ernst nehmen...
> [HWD, 34; emphasis added]

Heidegger does not just set out on his search by sampling random snatches of Hölderlin's words. The texts should illuminate what is most important about poetry. The sign of their centrality is their ability to force a decision. By setting as the goal of his search that which compels a decision from us, Heidegger indicates the seriousness with which he takes poetry. To do justice to his reading, we too must take it just as seriously. And yet, there is much about it that makes this essay difficult to take seriously.

"Hölderlin und das Wesen der Dichtung" is as widely-known as it is frequently repudiated. Heidegger obviously engages in an activity that does not accord well with standard approaches to Hölderlin's poetry. He is accused of distorting Hölderlin's words to serve his own purposes, of taking them out of context, of reinterpreting them when they seem to contradict his reading, of ignoring inconvenient alternatives to his

reading, among other things. Even writers sympathetic to his position are sometimes compelled to criticize his disregard for philology.[3] Heidegger has been accused of turning Hölderlin into a mere "mouthpiece" for his own ideas, so that the poet can be made to speak what Heidegger wishes to hear.[4]

And yet, it is only natural for Hölderlin to sound like Heidegger, since Heidegger does the speaking. Heidegger speaks, but with Hölderlin's words. He does not simply appropriate Hölderlin's words; however, they must assist him in speaking with the poet. Heidegger engages in a "dialogue" with Hölderlin, and this is our third "leading word." A dialogue can be valuable by providing an opportunity for immediate response from the other partner, unlike a monologue, in which only one side can speak. Both parties must share the right to speak and the duty to listen. Here, however, Heidegger does not seem either to let Hölderlin speak or to let the words "speak for themselves." In fact, however, Hölderlin has both the first and the last word. He has the first word, because it is his word which provides the starting point of the dialogue. Each of the "leading words" finds its origin in a text of Hölderlin's. He also has the last word, because ultimately his words will show how well or poorly Heidegger has listened to him.

At the same time, of course, Heidegger does not just listen, but comes prepared with his own issues. He is not a *tabula rasa*, but a thinker looking to understand poetry. His own pre-understanding will determine what he finds. Instead of attempting to deny or neutralize this implication, he accepts it and attempts to use it towards his goal. Just as his pre-understanding will affect the results of his reading, so will his reading affect Heidegger himself.[5] Heidegger thus comes with a question (*"Warum ist Hölderlins Werk gewählt?"*) and Hölderlin answers, at least as far as Heidegger lets him. It appears that Heidegger has the last word: he can shut off Hölderlin any time he pleases. Indeed, and so it is in a dialogue. One talks and listens and the dialogue develops—until one hears the answer to one's questions. Ultimately, it does not matter which of the partners in the dialogue *supplies* the answer.

How can Heidegger have a "dialogue" with Hölderlin, when the poet cannot answer, having died almost a century before *"Hölderlin und das Wesen der Dichtung"* appeared? Was Hölderlin not defenseless against the misuse or misunderstanding of his words? Actually, such an asymmetry does not impede the dialogue. A dialogue with a living partner would indeed sound different, but its form would remain the same. As we shall see, Heidegger has a particular idea about the nature and function of dialogue. The dialogue serves to bring speakers together on something. Does Heidegger's dialogue with Hölderlin do this? I believe it does.

In order to see one thing clearly, everything else must be relegated to the background. Heidegger shuts out much. He does not see the meters and rhythms of Hölderlin's verse; he does not see Hölderlin as a man of the eighteenth century with an idiosyncratic personality and specific friendships, concerns, and problems. He does not see Hölderlin as politically engaged. He does not even see or discuss the many mythological references in Hölderlin's writing. What does Heidegger see? "It" cannot be reduced to a word or concept; "it" is rather temporal. "It" must be found through a reading of his text. In the end, Heidegger reads the *Wesen* of Hölderlin's poetry as an *Anwesen*, as something taking place, a process—not a thing (*viz.*, an "it").

Although later commentators on Heidegger's essay have occasionally grasped this fact, they have seldom done it justice. They have been satisfied with packing it into a formulation similar to the one above, which inevitably distorts Heidegger's insight. By placing his thought into such a framework they reify it, they make it a commodity that can bartered or hoarded. It ceases to be a thinking and becomes merely another thought.

I propose to follow Heidegger's thinking through his encounter with Hölderlin. I am interested in the temporal process each engages in. But how can I "follow" Heidegger at all? Am I to "summarize" him, or "explain," or "apply" him? None of these; I am entering into a dialogue with him, just as he has entered into one with Hölderlin. We shall see how important the dialogue is, but for now we must be content with observing its beginning.

Heidegger's "Introduction"

"Hölderlin und das Wesen der Dichtung" begins with what seems a bald assertion: *"Hölderlin und er allein [ist] gewählt."* And yet, this is not quite the beginning. Before it comes a question: *"Warum ist ... Hölderlins Werk gewählt?"* As this question shows, Heidegger is quite aware of the peculiar nature of his title. Was it not quite naïve, in the fourth decade of the twentieth century, to claim to be talking about the

"essence" of anything? Besides, does not *Wesen* (i.e., *essentia*, "a fixed and timeless possibility of existence"[6]) belong to a metaphysical tradition Heidegger claims to have left behind? Nor does he explain how he means the term; indeed, he insists on seeking *"das Wesentliche des Wesens,"* which only seems to obfuscate matters. All of these questions, going unanswered by any explicit definition, force the reader to pay careful attention to the *use* of the word *"Wesen"* as the only way left open to discern its meaning here. The method of reading becomes itself the guide to the meaning of what is read.

Further, the choice of Hölderlin cannot help but seem personally tinged by the particular circumstances of Heidegger's place and time: Germany in the 1930s, a period of intense nationalism. In the National Socialist state, literature was to be purged of decadence and poets like Hölderlin and Kleist were refurbished as proto-fascists.[7] Under such circumstances, and given Hölderlin's shared geographical proximity, it would seem easy to understand why Heidegger chose him and not Dante or Shakespeare to speak for poetry. Heidegger would seem to be interested in Hölderlin because of a personal affinity to the poet. Heidegger would seem to be nominating Hölderlin as the poet of the German *Volk*. The lack of any attempt to "justify" his choice of would make it appear that the choice rests solely on Heidegger's subjective preference and that he intends to monopolize discussion of poetry by removing Hölderlin from any comparison with other poets.

The problem with a reading of Heidegger as outlined above is that it holds him to standards of scholarship which he himself shuns in no uncertain terms. His objections to "scientific" philology were anticipated in *Sein und Zeit* (1927) and elaborated, as we have seen, in *"Der Ursprung des Kunstwerkes"* (1935–6). The charge of subjectivism is also misleading, since Heidegger in all of his writings repeatedly insists that being a "subject" is a privative way of being. It may be adequate for dealing with "objects" in an analytic or scientific study, but not for describing an encounter with a work of art. With its scientific pretensions, literary criticism similarly reduces its texts to "objects" which can then be analyzed, ordered, compared, and placed in a hierarchy. Following such an approach, we could say that Hölderlin's texts belong in the classical-romantic period of German literature, that they show discrepancies from the norms prevelant in those times, that they treat themes such as the relation of classical Greece to modern Europe, the crisis in belief systems, the importance of subjectivity, etc. By treating

Hölderlin's texts as *gleichgültig* with other texts, however, philology ignores their true importance. But Heidegger, as we shall see, far from using the concept of the *Volk* to establish the correctness of a certain version of Hölderlin's thought, actually uses Hölderlin's words to challenge and even *define* the German *Volk* itself. By responding to this poet's words, by accepting the challenge he presents them, the people coalesces and discovers its destiny.

Heidegger's essay thus begins with an explicit questioning of the radical implication of its title. The claim to discuss the "essence of poetry" is questioned by the author himself in his opening sentence. He has switched places with his audience and has questioned his own formulation. Nor is his question and reply merely a rhetorical flourish. Certainly, many people would turn first to Dante, Shakespeare, or Homer for world-class poetry. This being granted, Heidegger nevertheless re-affirms Hölderlin's priority in the search for the *essence* of poetry. He then introduces another question (or objection), followed by supporting statements, and a response to them. Through this procedure of question-and-response, Heidegger gradually makes clear what he means by the *Wesen* of the title, which is such that a comparative philological approach could never hope to find it. He does not proceed linearly from assertion to proof, nor from particulars to generalization. It is difficult to say where exactly he *does* begin: is it with the title? or with the five *Leitworte*? or with the first sentence? or with the "first" section (which itself is preceded by "1.")?

This initial ambiguity, which is only too easily read over, actually belongs integrally to Heidegger's method. He begins with a facile understanding of the object before him. That understanding brings up questions which require another look at what has already been said. In such a hermeneutic process, every possible starting-point is formed in advance by a pre-understanding (*Vorverständnis*) which itself gets modified by interaction with the matter to be understood (in this case, a literary text). Understanding is thus a process of alternating approach/withdrawal/return. Each step in the process is only important as a transition to another step, while ultimately returning to the beginning. This hermeneutic process can also be found within the *Machtbereich* of the word *Feiertag*. The final chapter of this study will return to listen again to this word within its context in the poem *"Wie wenn am*

Feiertage...," but for now we must be content if we can learn to understand Heidegger's hermeneutics.

In addition, Heidegger does not shrink from contradicting himself, usually by bringing out a previously unexamined ambiguity in the very words used to frame the discussion. Contradiction is not the stopping point but the beginning of the interpretive process. Indeed, one might go so far as to say that before a contradiction arises hermeneutics is not needed.

Heidegger elucidates his concept of *Wesen* by setting it off against the commonly understood concept of the "general essence of poetry." Such an understanding has a long history in philosophy. Heidegger does not stop to argue about it, but points to its "indifference " (it is *"das Gleichgültige"*). Such an essence can never become essential: it allows "subjects" to remain complacent in their contemplation of "objects" of discourse, without ever needing to recognize that a *decision* about those objects and about the observers themselves has already been made. Heidegger seeks that elusive "something" that *"forces us to decide if and how we will take poetry seriously in the future...[that which] enables us to stand under the sway of poetry."* [HWD,34] Unlike the philosopher or philologist, who can calmly analyse the nature of poetry without drawing an explicit connection with his or her own life, Heidegger recognizes the radical nature of the call which poetry makes to its audience. Poetry does not tolerate half-answers. Perhaps it is worthless, a distraction, without meaning for life; then we must say so and leave it as an occupation for idle moments. But if, on the other hand, poetry is important to us, we must have the courage to admit that and draw its conclusions. We cannot shelter ourselves from its claim on us by labelling poetry "mere words." Words are not simply found in our environment and picked up or dropped on a whim. They "belong" to man; they are either "my" words or they are "not-mine." If they truly belong to me (i.e., if they "speak" to me), I cannot choose to ignore them. To hear my name called, even if I do not recognize the caller or where the call originates from, calls me to myself in a way that cannot be refused. Only after that call of poetry has been heard can anything "essential" be said about it. The necessity and danger of dialogue is thus opened.

Heidegger, in choosing Hölderlin, has chosen "five leading words." They are "leading" because they do not remain encapsulated within themselves, but point beyond themselves to something more essential.

They are thus like "leading" questions, which imply more than they say. He does not discuss the reason for choosing these particular "words" before all others that Hölderlin wrote, just as he did not justify choosing Hölderlin above other poets. The five leading words are not the only possible entry points to the essence of Hölderlin's poetry. Nor do they seem to be hierarchically organized; i.e., they do not lead to a conclusion, nor do they spring like examples or consequences from a unitary, fundamental insight. Nor does Hölderlin himself seem to have intended them as apodictic truths. Are they then arbitrary? Not if judged by their significance for Heidegger's goal: thinking the essence of poetry. Words used in argumentation look backwards, to the source of their authority, which they then usurp. Used hermeneutically, words can only be valuable if they look forward and thus facilitate the process of understanding. Heidegger is not looking for the rightful successor to the "Word of God" but for the best guide to unexplored territory.

The Five "Leading Words"

1.

[das Dichten:] "diss unschuldigste aller Geschäffte"

What does the "poet of the poet" himself say about poetry? In these words from a letter to his mother he affirms its harmlessness and innocence. How is poetry "innocent?" It is free from the trap of causal thinking which attends all pragmatic decisions. Poetry does not directly affect behavior. It thus exists in a realm of freedom. Poets are not held to the same standards that apply to historians or philosophers. The price for this freedom, however, is that poetry does not take part in the world of action. It is powerless to stop a bullet or a bad cause. Its powerlessness, however, frees it from any taint of guilt. Heidegger suggests that guilt only arises within the realm of action and, since poetry is like a "game of words"—i.e., distinct from immediate action—it is the "most innocent" of occupations. He emphasizes at length poetry's harmlessness, inconsequentiality, and dreaminess.

Of course, looking at poetry this way, while seeming the most direct and natural approach, will never lead to its essence, which is Heidegger's stated goal. Many people will take poetry as basically just a game and decide *against* it being anything more. Thus its innocence will have captured them, entrancing them in their own innocence (i.e., naïveté) into accepting its surface as essentially transparent. Heidegger's hermeneutic calls for the courage to start with the naïve, the willingness to be taken in, at least for a time, by the "mere appearance" of a thing. To resist the superficial would be to remain within the realm of the superficial, in which one must constantly be on guard against charges of superficiality. He begins, as in *Sein und Zeit*, with that which appears *zunächst und zumeist*, i.e., the average and everyday appearance of the phenomenon. Having gained an opening into the essence or nature of poetry through its public appearance, he must then leave it behind.

Poetry, in its everydayness, *is* a game; its formalized rules, meters, tropes, even its "occasions," afford much pleasure and offer themselves in a multitude of combinations. But this does not really tell us much about poetry. Once we have recognized poetry's freedom as a game, however, we are led to consider its origin. Games are free because they are rule-bound: they have distinct boundaries and goals, which separate them from everyday reality, and which require language in order to exist. Language is the field in which poetry can mean something and in which it can also mean many things. As soon as language permits communication, it also makes possible miscommunication. That is, words may be used not only to inform, but also to conceal or mislead. Thus our investigation of the superficial side of poetry has led us, perhaps surprisingly, to its fundamental ambiguity.

I feel tempted to call Heidegger's exposition of poetry as a game "ironic," so exaggerated is his choice of descriptors. Each statement about poetry is in fact a well-worn cliché. But irony would be out of place here, because even these clichés have a share of truth to them, which Heidegger does not dispute. Their indispensible service is the open and clear way they contradict what most lovers of poetry would say about it.

Heidegger himself has already contradicted implicitly his initial statement that the essential element of the essence of poetry is *"jenes, was uns zur* E n t s c h e i d u n g *zwingt..."* [HWD,34, emphasis added] Now he declares that poetry

> ...entzieht sich damit dem ernsten *Bereich der Entscheidungen*, die sich jederzeit so oder so schuldig machen. [HWD, 35; emphasis added]

Poetry thus forces decision while itself eluding the field of decisions. It is innocent, since it is free from consequences in the world of action, and yet—somehow—it essentially forces man to decision. How can we hope to find something so immediate, so absolute that it forces such a confrontation, in an occupation that seems so impotent? It is hard to believe that Heidegger has set out on the right path, if his goal is to find such an essential poetry. We find ourselves in a dilemma, since neither side of poetry can be jettisoned without disfiguring what we know or feel about poetry.

Is poetry then deceptive, as Plato said? How can that which lays the foundation for man's Being (in language) be deceptive? And yet poetry does deceive, just as truth as unconcealedness deceives.[8] Poetry presents itself as harmless and yet is anything but harmless. It must protect itself from being appropriated by instrumental language by appearing as a mere "game." On the other hand, Hölderlin's use of common words, mythological and biblical figures, gnomic sayings, large poetic forms, and frequent intratextual allusions all contribute to the deception that there must *really be* some underlying structure, goal, or meaning in his work. Surely the great majority of readers have at some time or other accepted this belief; many books have been written with the purpose of *establishing* just such a fiction. The playfulness of poetry has disarmed them and they in turn have sought to disarm it. It is only after having been seduced by the innocence of its surface that one comes to face the seriousness and complexity of Hölderlin's poetry.

Heidegger continues throughout the essay to proceed in this fashion: advancing a statement, questioning it until the contradictions inherent in the relationship of question and answer give rise to a different question and thus to a different possible answer. In this case, the contradictions between the nature of poetry as we ordinarily think of it and as it appears at special moments (at a wedding or a funeral, say, or when one is in love), lead us to look at poetry all over again, to start from the ground up, so to speak. Heidegger suggests here explicitly that poetry uses language as the "material" (*Stoff*) out of which it creates its works of art. Such a view, taken as his final, ultimate *thesis* about language, would again trespass into the cliché-ridden generalization, as the statements referred to above. It would also contradict what we have seen Heidegger arguing at length in *"Der Ursprung des Kunstwerkes."* The dichotomy of "form" and "matter" has showed itself to be essentially inadequate to grasping an artwork originarily. Instead, its use here is meant only to serve as a starting point for further thinking. Poetry may be innocent, but it grows out of the soil of language, *"the most dangerous of goods..."*

2.

"..Darum ist der Güter Gefährlichstes, die Sprache dem Menschen gegeben, ... damit er zeuge, was er sei ..."

Structurally, this section of the lecture also begins with a step forward and a step back. It steps forward to answer the question posed in the last sentence of the previous section, *"Was sagt Hölderlin über die Sprache?"* A quotation is introduced which describes language as "the most dangerous of goods." The contradiction with the previous interpretation of language is noted briefly, but then set aside. The discussion steps back to consider three *Vorfragen*, each of which requires a closer, slower, more "timely" look at the terms in which the answer to the first question was posed: *Mensch*, *Gut*, and *Gefährlichstes*. Through the process of this looking the contradiction takes on new meaning, which, however, does not fully appear until the end of the essay.

What has so far emerged from a structural reading of *"Hölderlin und das Wesen der Dichtung"* is the basic movement of Heidegger's process of interpretation. After beginning with a "questionable" statement and its steadfast reaffirmation, Heidegger has introduced another statement—seemingly innocent and uncontroversial—but has shown how it too can be questioned and how it can lead to yet more fundamental questions. Each apparently uncontroversial assertion contrasts starkly with its predecessor. But the expected reconciliation does not take place; rather, the whole problem is "set aside." Before the contradiction between "innocent" poetry and "dangerous" language can be worked out, each side must be better understood. The working-out of this contradiction frames the remaining discussion.

The crux of the contradiction, in my opinion, is found in poetry's demand for a decision. In the previous section Heidegger seemed to remove poetry from the realm of decision. Now he reintroduces poetry into this very area by listening carefully to Hölderlin's word *zeugen*.

In everyday language, *zeugen* means "to bring forth" in one of two senses: either 1) by producing something, whether an object or a person, as in "procreation," or 2) giving evidence by stepping forward in public and speaking about what one has personally witnessed.[9] It is the second sense which Heidegger takes up. The value of one's *Zeugnis* depends upon one's perceived value as a person; i.e., one's trustworthiness. Testimony testifies as much about oneself as about another. *Zeugen* is thus not just one activity among others, to be taken up as circumstances require, but is rooted directly in the kind of being man is. To fully understand the importance of *zeugen*, we must consider how Heidegger views man (Dasein).

In *Sein und Zeit*, Heidegger argues against the scientific view of man, which he claims falsifies man's true nature. Human beings do not "have" Being like one has a pulse or brain activity. Rather, the Being of Dasein is always incomplete, it is had as something *to be*. *"Das 'Wesen' dieses Seienden [Daseins] liegt in seinem Zu-sein."*[10] One cannot "sum up" a human life; it is always "in process," and it always lacks something which must be filled in. Later in the same book, he puts it that *"in Dasein there is always something* still outstanding*..."* [B&T 279, emphasis added]. That which "stands out" (*noch aussteht*) is the potentiality-for-being that Dasein fills in by living in a certain way and not another. Put another way, the issue at stake in life only gets worked out through living, and, more specifically, in terms of *how* life is lived. It is not a matter here of the *content* of the way chosen but of the *act* of choosing as determinative of Dasein's Being.

> Dasein always understands itself in terms of its existence—in terms of a possibility of itself: to be itself or not itself. Dasein has either chosen these possibilities itself, or got itself into them, or grown up in them already.
>
> [B&T, 33]

Heidegger is not passing judgement on different "kinds" of Dasein, but rather describing Dasein's constitution. Dasein always understands itself somehow, and that understanding makes certain things possible for it, while other things are "out of the question." In either case, choices are made or accepted on the basis of Dasein's Being-its-Self. By following up or neglecting these possibilities, Dasein discovers itself in the world. It must be emphasized that the "self" Dasein discovers this way is not ontologically privileged; on the contrary, it is precisely *not* "itself."[11]

In his book on *Being and Time*, Hubert Dreyfus lays emphasis on *"Dasein's special way of existence,"* which is not a kind of "thing" (*existentia*), but an ongoing process of self-interpretation:

> To exist is to take a stand on what is essential about one's being and to be defined by that stand. Thus, Dasein is what it interprets itself to be.[12]

"Taking a stand" on one's Being is what Heidegger sees in Hölderlin's word *zeugen*. Earlier in this essay Heidegger put it this way: *"Doch eben dieses Wesentliche des Wesens suchen wir, jenes, was uns zur Entscheidung zwingt..."* The essence of such a *Wesen* is its demand for a decision. Only after that decision is "made"[13] (whether or not it is recognized) can something "matter" and thus become essential. Hölderlin's words quoted as a *Leitwort* lay similar weight on testifying to one's existence. Man is that being who must give evidence (*zeugen*) of what he is. That is, "giving evidence" or "making a statement" (and thus language) belong essentially to man. One bears witness to one's own existence by stepping forward into the "open" space (*das Offene*).

According to Heidegger, one who bears witness "stands in" for something, *viz.*, his statement. Without vouching for the truth of what one has said, one cannot be a witness. *"Der Mensch ist der, der er ist, eben in der Bezeugung des eigenen Daseins."* [HWD, 36] Human beings are capable of many types of action, but these are all "immaterial" except for one thing. Man only comes to be in the witnessing of his own existence. This does not refer to a subjective recollection or an objective speaking out; rather, by standing in for one's own existence, one *thereby* creates that existence. The important point for Heidegger, and I believe also for Hölderlin, is that such witnessing is ineluctably linguistic in nature. Only in language can one bear witness to an understanding of what it means to be human.

Even more significantly, *zeugen* does not simply imply linguisticality. To bear witness to his existence man must also "stand in" for what he has witnessed. Unlike mere linguistic utterances ("words"), which can change or vanish in a moment, the *Zeugnis* derives its importance from its foundational relation to the "witness" and the "witnessed." *Zeugen*, thus, is connected to the necessity for decision which characterizes the essence of poetry.

> Die Bezeugung des Menschseins und damit sein eigentlicher Vollzug geschieht aus der Freiheit der Entscheidung. Diese ergreift das Notwendige und stellt sich in die Bindung eines höchsten Anspruchs. [HWD,36]

For what does man stand in? For his "belongingness" to the earth. He bears witness in everything he is and does to the particular "earth" from which he has sprung—not in the sense of a region, or country, or landscape, but as the past which he himself is. Dasein always finds itself in its "thrownness," (*Geworfenheit*) or facticity. As such, it finds itself as "having-been:" *"Nur sofern Dasein überhaupt* ist *als ich* bin*-gewesen, kann es zukünftig auf sich selbst so zukommen, daß es* zurück*kommt."*[14] But the things of the earth are in constant battle and contradiction with each other. To stand in for the earth is thus to bear witness to the *Widerstreit* of things: Heidegger asserts that this is what Hölderlin means by *Innigkeit*. Here he is referring to the lines just before those chosen as *Leitwort*, in which Hölderlin says, *"Aber in Hütten wohnet der Mensch, und hüllet sich ins verschämte Gewand, denn inniger ist/achtsamer auch und daß er bewahre den Geist ... diss ist sein Verstand."* [HWD,35]

It might seem that *Innigkeit* refers to the state of having an inside, of subjectivity, of "emotions" as opposed to "actions," etc. In that case, these things could also be lacking and thus they would not be essential to man. Heidegger does not discuss it from this angle at all. *Innigkeit* gives evidence of the conflict of things. Once we recognize multiplicity, we are forced to acknowledge conflict among those things that are "within" the world. If there were an "external" standpoint from which man could have access to the world, the possibility of harmony would not be ruled out. No matter how much the things of the world might conflict among themselves, they would not penetrate to man-the-observer. Such a standpoint is attempted and often achieved by science. Since historical man is, on the contrary, thoroughly involved in the world, which is *his* world, he is always already caught up in the conflict. He has effectively "chosen sides" as soon as he discovers himself in this ("my") world.

Hölderlin, as I read him, seems to be saying that man is *inniger* because he relates to his world through understanding (*"...daß er bewahre den Geist, ... diss ist sein Verstand"* [HWD,35]). Man not only exists "inside" but *knows* that he exists inside. For that reason, he also feels shame and hides himself in clothing and huts. "Understanding" here is perhaps understood as Aristotle's *logon echon*: thus the reference to language (*logos*). By using his understanding, man keeps things apart and gives them their identities, as well as bringing them together in his *Innigkeit*. Even destruction is a way of giving evidence to a world, being joined through mutual opposition to creation. Heidegger takes Hölderlin's word *Innigkeit* out of the field of subjective reflection and restores its more existential meaning. "Innerness" for Heidegger describes man's Being-in-the-World.

> Die Bezeugung des Zugehörens zu dieser Innigkeit geschieht durch das Schaffen einer Welt und ihren Aufgang ebenso wie durch die Zerstörung derselben und den Untergang. [HWD,36]

Language is "the most dangerous good" because it brings forth the whole world of beings, thereby making it possible to mistake them for Being itself. *"Gefahr ist Bedrohung des Seins durch Seiendes."* Language creates the bare possibility of deception and disappointment: only after things have names can they be wrongly named or forgotten in favor of their names. The real danger, though, is not that one being will be mistaken for another, but that Being will be mistaken for just another kind of being. Such thinking misleads us into the error of thinking that we can actually give an answer to the question of what it means "to be." Living in the anonymous world of average everydayness (*das Man*) always threatens to absorb our attention totally. It seems so easy to just do what is expected and thus to forget what is most essential. In average everydayness, life is lived along the lines set for "anyone." To remain within that world, we learn to exclude from our speech anything which cannot be "commonly understood." We thereby forget how to speak about what cannot be commonly understood, what demands the utmost individuality from Dasein. Authentic language does not simply communicate, it always occurs as a happening (*ein Ereignis*). Clearly, there is something temporal here.

But what is meant by "the most dangerous good?" We have seen how "essentially" dangerous language is. When we speak of something as a "good," we often think of it as a possession, a means to some end apart from which it exists independently. Language can certainly be seen that way, as a means to communication, for example. Such a conception, however, completely misses the importance for man of bearing witness to his *Zugehörigkeit zur Erde*. Language first opens the possibility of standing in the openness of beings. There could be no world, there could be no witnessing, without language to create the possibility of meaning for man.

> Nur wo Sprache, da ist Welt: das heißt: der stets sich wandelnde Umkreis von Entscheidung und Werk, von Tat und Verantwortung, aber auch von Willkür und Lärm, Verfall und Verwirrung. [HWD,38]

"World" is here understood as the ever-recurring circle of decision, deed, and responsibility. Heidegger asserts that language provides a guarantee to man that he can exist as a historical being. *"Nur wo Welt waltet, da ist Geschichte."* [HWD,38] This means that man, through language, has access to history as a realm in which he can exist, even though it is not "present" to him in strictly spatio-temporal terms. Language ceases to be merely a tool and occurs as a truly fundamental event, accessible only to human beings. Heidegger does not assert or imply that all people live historically at all times, but that through language the possibility of entering historical discourse is given to humans.

3.
*"...Seit ein Gespräch wir sind
Und hören können voneinander."*

Developing the idea of language as more than merely words and rules of usage, Heidegger discovers its true origin in dialogue (*Gespräch*). A positivistic bias seems to have led many people to assume that language must have existed prior to any "dialogue," just as paints must exist before a painting can come to be. Such a view reduces language to isolated words, sounds, marks, etc., and then imagines someone who comes along and combines them in a particular order and exchanges utterances with another human being. It handles phonetics and phonemics very well, but it is awkward in semantics.

In contrast to this view, Heidegger says that language only comes into being through dialogue—just as man's existence only comes to light through *zeugen*—and that dialogue itself is dependent not on speaking, but on the ability to *hear.* Does there not have to be something said before anything can be heard? Is there no temporal priority of producer to thing produced? Indeed, but only if language and speaker are taken as separate from each other. What if the producer produced himself? What if language produced the being characterized as language-bearing (*logon echon*)? We must consider that language, if taken as a collection of sounds arranged according to rules for purposes of communication, could be (and has been) employed by any number of animals while lacking precisely that which makes it central to humans. It seems clear that what truly differentiates language as human speech from language as arbitrary rules and sounds, is the shared understanding coming from dwelling together in a "world." What Heidegger says about "discourse" (*Rede*) in *Being and Time* applies here to his remarks on language:[15]

> Mitdasein ist wesenhaft schon offenbar in der Mitbefindlichkeit und im Mitverstehen. Das Mitsein wird in der Rede "ausdrücklich" ***geteilt***, das heißt es *ist* schon, nur ungeteilt als nicht ergriffenes und zugeeignetes.
>
> [SuZ,162; italics in original]

> Dasein-with is already essentially manifest in a co-situatedness and a co-understanding. In discourse Being-with becomes 'explicitly' ***shared***; that is to say, it is already, but it is unshared as something that has not been taken hold of and appropriated. [B&T,204, slightly altered]

Not a particular kind of sound, but a kind of hearing, is essential for the phenomenon "language." It only degenerates into an inventory of words, phrases, rules, sounds, etc., when one excludes from consideration (forgets) its function of bringing together (*"sich gemein machen"*).

Heidegger, however, does not forget. In order to remember how language can be the basis of community, he next introduces the concept of the "dialogue" (*das Gespräch*). Already from a glance at the etymology of the word, *das Gespräch* < *sprechen*, a bias in favor of speaking appears. Ordinarily, we think of a dialogue as two people alternately speaking and listening to each other. In this situation, however, the preferred position is clearly that of the speaker, and the dialogue itself only becomes important as a human activity because it explicitly allows both parties to *speak*. Heidegger does not look at it that way. He has Hölderlin's word leading him to the complementary and even more fundamental requirement of *hearing* each other:

> ... seit ein Gespräch wir sind
> Und hören können voneinander. [*"Andenken"*]

Heidegger here uncovers the importance of the dialogue by another of his famous reversals: it is not the ability to speak which is essential for dialogue, but the ability to *listen*. *"Das Hörenkönnen ist nicht erst eine Folge des Miteinandersprechens, sondern eher umgekehrt die Voraussetzung dafür."* And yet, in order to be able to hear anything in the dialogue, words (and thus speaking) must also be present. Both are equally essential to the hermeneutics of the dialogue. Further, humans do not "have" a dialogue, according to Hölderlin; rather, they "are" a dialogue—i.e., their Being consists of dialogue—and indeed it is a single, unified dialogue. *"Das Gespräch und seine Einheit trägt unser Dasein."* [HWD,39] Heidegger emphasizes that the ability to hear one another in dialogue presumes the unity of what is spoken and heard. *"Wir sind* ein *Gespräch."* He is not trying to short-circuit the problem of meaning in this way. He is not concerned with whether you hear what I say—i.e., if you understand my words the way I intend them. We must remember that this dialogue is no ordinary dialogue. It takes place between men and gods. It is consitutional of their relationship to each other and is unavoidable, even when unrecognized. It brings men and gods together and thus unifies them.

When dialogue occurs, speaking *and* hearing are shared. *"Dabei vermittelt dann das Sprechen das Zueinanderkommen."* [HWD,39] Dialogue brings them together in a unified, lasting, presence which is common to both.[16] Just as speaking without hearing is impossible, so too do speaker and hearer belong as much to each other as to themselves. The shared meaning of the dialogue (*das Eine und Selbe*) is "*das wesentliche Wort*." Only by agreeing to step into the common arena of the dialogue can language speak "essentially." To succeed, the dialogue must bring out something lasting, about which one can actually *converse*. The discovery of permanence or stability presumes the prior existence of temporality and transience. Only after "raging time" has begun can something count as "enduring." Time, springing from language in the dialogue, makes historical being a possibility for man. *"Seitdem die Zeit aufgestanden und zum Stehen gebracht ist, seitdem* sind *wir geschichtlich."* [HWD,39]

Of what does authentic dialogue consist? Is any exchange of words a conversation? That would be similar to the language created by computers: a purely formal and empty definition. But since language first makes possible the very being "man," it cannot be empty of meaning. Indeed, Heidegger shows that language alone creates the possibility of "meaningfulness" and "worldhood." The word for the only beings capable of creating a world autonomously has always been: the gods. Thus, the gods and language are coeval. But, since man himself is the language-bearer par excellence, then he must the one who calls the gods by naming them. By coming together in dialogue they in turn bring forth that being subsequently known as "man:"

> ...daß im Nennen der Götter und im Wort-Werden der Welt gerade das eigentliche Gespräch besteht, das wir selbst sind. [HWD,40]

The gods can only appear in language when they do the calling and we respond to that call. That is, by naming the gods (e.g., in myth or in holy scriptures), man calls them to approach and enter into dialogue with him. But such a calling has already responded to an demand (albeit unspoken) by the gods themselves: man is called to take responsibility for his own destiny (*Schicksal*), to make it his own. When the gods bring man's existence into speech, we are brought before the necessity of a decision as to whether we will speak with them or without them.

> Indem die Götter unser Dasein zur Sprache bringen, rücken wir erst ein in den Bereich der Entscheidung darüber, ob wir uns den Göttern zusagen oder ob wir uns ihnen versagen. [HWD,40]
>
> Only when the gods bring our existence into language do we then enter the realm of the decision as to whether we will address ourselves to the gods or whether we will deny ourselves to them.

The gods, in their dialogue with men, bring out the decision on which a historical people bases itself. Heidegger is not saying here that a people deliberately sets out to "decide" whether or not to have gods and, if so, what kind to have. The essential thing about gods is that one does not "have" them: either they are simply "there" or they are absent. Just as Dasein, understood individually, has always "taken a stand" on itself, so too does a people or culture. Hubert Dreyfus has formulated this difficult concept in the following way: *"Each Dasein must understand itself within some culture that has already decided on specific possible ways to be human—on what human beings essentially are."*[17] Dasein does not find itself alone; it is in a world which it shares with others, a culture. Every people sets up its gods, although it does not always display or even recognize them. The gods are clearly not meant as just an anthropomorphism, but rather symbolize the highest ideals of a culture. Thus, when a people *"den Göttern ... versagen,"* it fails to live up to the mark it itself has set. Even when a people repudiates all gods, it remains within the dialogue of gods and men. According to Heidegger, a decision has always been "called for" with respect to the gods.

> Seit die Götter uns in das Gespräch bringen, seit der Zeit ist es die Zeit, seitdem ist der Grund unseres Daseins ein Gespräch. [HWD,40]

The dialogue thus gives man in true language the basis (*Grund*) of his existence. Whenever talk turns to the most fundamental things, the gods have been brought in, even if only to be denied. Heidegger next turns to consider how this talk begins. Someone, after all, must first utter the words which name the gods and name us to them.

4.
"Was bleibet aber, stiften die Dichter."

Man cannot speak "the essential word," that which is "one and the same" (*das Eine und Selbe* [HWD,39]), without something on which both speaker and hearer can agree. But how can something so stable and "*agree*able" be found? Can science, can philosophy bring it somehow out of chaos? No. Here is where the poets must come to our aid.

They found the lasting in and through their word. The poetic word which names the gods thereby gives them voice and creates the foundation for what can endure and maintain its value. Heidegger seems to be thinking of Homer (possibly also of Moses), whose stories "name" the gods to their identities and make possible a common frame of reference for his people. That is, only through Homer do we hear the gods speak, discover their identities and demands and thus learn the ultimate consititution of the world. Of course, Homer himself was only one, and not even surely the last, of a whole tradition of poet-singers of the Trojan War. Yet—somehow—his word resonated so strongly with those who heard it that it could not be forgotten; indeed, it demanded to be repeated and committed to memory. Gerald Bruns captures this aspect of Heidegger's thought nicely:

> [T]he poet names all things, but such naming doesn't merely designate what is already there, rather it brings things into appearance, establishes them for the first time as being what they are. ... Poetry is not about reality; it is what gives us reality.[18]

Only the poetic word continues to be spoken and thus remains alive, essential. The enduring is not something always present-at-hand, always available for man's purposes but existing apart from him. The enduring (*das Bleibende*) must be *achieved,* it itself must be established, in order to not simply pass away like everything else in the flux of time:

> Aber kann das Bleibende denn gestiftet werden? Ist es nicht das immer schon Vorhandene? Nein! Gerade das Bleibende muß gegen den Fortriß zum Stehen gebracht werden; das Einfache muß der Verwirrung abgerungen, das Maß dem Maßlosen vorgesetzt werden. [HWD,41]

Heidegger's rhetoric here is emphatic, again showing his use of the reversal. "The enduring" (taken as *substantia*) is not, as we would expect, the more basic concept, in terms of which all else must be measured, and which entails that time becomes a "problem" to be explained. Rather, time (or change) is the more fundamental concept—indeed, it is only after uncountable failures taking place over time that the very possibility of permanence emerges. Heidegger's use of the word *Stehen* also suggests not merely bringing something to a "standstill," but also bringing it to a *stand*, standing up, standing out and thus appearing for us. Permanence must be "brought to stand," simplicity must be "grappled away" from confusion, the stability of measure and law must be "brought forth" against chaos. The creation myths of so many peoples show time being wrested from a prior condition or "age" of chaos. Clearly, what requires effort to accomplish cannot be original or fundamental.

The poet says the essential word which first allows Being to appear. *"Dichtung ist worthafte Stiftung des Seins."* [HWD,41] Being can never be deduced from beings; it must be freely "created" (in the sense of *schöpfen*). Such creation is a giving (*Schenkung*) and a founding (*Stiftung*). We are reminded here of the discussion of poetry in *"Der Ursprung des Kunstwerkes,"* in which poetry was seen to have a threefold character of *Stiften*: *Schenken*, *Gründen*, and *Anfangen* [UK,61]. These are extraordinary terms for describing the poetic process. The essence of things must be "given" by the poet, in the manner of an a priori, but without the oppressive structure with which it is usually encrusted. Because the *Wesen* of things cannot be deduced or calculated, it must be accepted as a gift when it steps forward. The poet is the one who waits, listens, and takes up the gift when it is offered. By such seemingly insignificant behavior he or she is able to provide a new foundation for language and thus for Being.

The creative activity of the poet is a free giving, but it also establishes the foundation of all things. It takes part in the freedom and innocence of the game but at the same time it produces as a consequence the most fundamental of results. The root of the contradiction between the first two *Leitworte* now begins to appear more clearly. Poetry is the most innocent of pastimes, but acquires the responsibility of laying the basis for all subsequent meaning and thus for danger and guilt.

5.

"Voll Verdienst,. doch dichterisch wohnet Der Mnsch auf dieser Erde."

Man lives actively—he produces things, for example. He deserves full credit for his achievements (*Verdienst*). His deeds are his own. But human existence is not a single person's achievement; it is received as a gift and a fate. A gift, because man does not himself acquire it; a fate, because once having received it, he must himself work out its/his end. According to Heidegger, Hölderlin sees human life quite simply as "poetic." In using this term, though, we must be aware of its broader significance, as set out in *"Der Ursprung des Kunstwerkes."* "Living poetically" does not simply refer to an aesthetic awareness of language.

> "Dichterisch wohnen" heißt: in der Gegenwart der Götter stehen und betroffen sein von der Wesensnähe der Dinge. [HWD,42]

Being "struck" (*betroffen*) is not a gentle tap on the shoulder or a beautiful vision. Rather, it goes right through to the very ground of one's Being. The reference in this statement is surely to *"Wie wenn am Feiertage...,"* in which Semele is described as "*die göttlichgetroffene*," and where the poet's soul is said to be *"schnellbetroffen"* at the moment of poetic conception. No matter how great man's accomplishments may be, they do not suffice to reach the real depth of human existence, which, according to both Hölderlin and Heidegger, is "poetic."

Living poetically, man casts forth what has been given (*Geschenk/Geschick*). This casting-forth is the active and creative embrace of what has been handed over from the gods through the poets. Hölderlin's verses underscore through the contrastive *doch ("Voll Verdienst, doch dichterisch wohnet...")* man's inability to reach behind the gift of poetry (language) to found human life on itself. Rather, mankind takes up what has been handed down by the poets and thus "dwells poetically." In this passage the distinction between poet and people is not significant: *all* must take up what the gods have handed over to them, whether they stand near to or far from that source.

The essay began with the contradictory statements that poetry was "the most innocent of occupations" and yet it works within the realm of language, "the most dangerous of goods." At the time, we saw that Heidegger simply brought out the contradiction in Hölderlin's disparate words but did not attempt to resolve it. He has now arrived at the appropriate spot from which to explain how poetry can be both "innocent" and "dangerous." He does this by pointing to its foundational character in the threefold sense of *Stiften* which he elaborated in *"Der Ursprung des Kunstwerkes,"*[19] i.e., as granting, founding, and beginning:

> Dichtung ist das stiftende Nennen des Seins und des Wesens aller Dinge—kein beliebiges Sagen, sondern jenes, wodurch erst all das ins Offene tritt, was wir dann in der Alltagssprache bereden und verhandeln. [HWD,43]

Because it sets the stage for all that follows or can follow, poetry is both free (like a game) and dangerous. It is free because it is not determined by any other. It is dangerous because it is constrained to name Being only through beings, and thus the possibility arises that hearers will take the latter for the former. Thus, poetry is innocent *and* dangerous. These two contradictory thoughts must be capable of being thought together if the "essence of poetry" is ever to emerge.

But Heidegger does not stop there. Instead he questions the danger of poetry once more, answering this time from Hölderlin's letters, in which the poet claimed to have been struck by "Apollo." Heidegger comments:

> Die übergroße Helle hat den Dichter in das Dunkel gestoßen. Bedarf es noch weiterer Zeugnisse für die höchste Gefährlichkeit seines 'Geschäfftes'? [HWD, 44]

It might seem as if Heidegger were drawing on a psychological argument: poetry is dangerous because it may lead to madness (e.g., Hölderlin's *Umnachtung*). But it is the tension between the clarity and brightness (*Helle*) of poetry and the darkness into which the poet flees that determines this thought. Poetry brings the poet into immediate and dangerous contact with the source of Being. In order to withstand such presence, the poet requires the appearance of harmlessness.

The innocence of poetry makes it resemble a game, but it is no mere game, some harmless amusement (*Zeitvertreib*). In playing, people can easily come to forget who they really are, they can seek to hide from their own destiny, which no one else can fulfill. While poetry brings people together as in a game, its playfulness is such that each individual comes to stand on the ground of his or her own Being. Such a standing is a resting or remaining against the destructiveness of time (see above, 121) and thus far from a mere quiescence.

> Dichtung sieht aus wie ein Spiel und ist es doch nicht. Das Spiel bringt zwar die Menschen zusammen, aber so, daß dabei jeder gerade sich vergißt. In der Dichtung dagegen wird der Mensch gesammelt auf den Grund seines Daseins. Er kommt darin zur Ruhe; freilich nicht zur Scheinruhe der Untätigkeit und Gedankenleere, sondern zu jener unendlichen Ruhe, in der alle Kräfte und Bezüge regsam sind. [HWD,44]

Such a "rest" is not mere passivity, but rather the stillness and freedom of the "original spring" (*Ursprung*) out of which all power originates. Collected on the ground of his Being, the poet rests with a power that calls together the people and the gods. This is his freedom.

Heidegger goes on to show that poetry's freedom does not prevent it from also being "bound," and in fact in two ways. It is bound to the gods as well as to the people. Poetry is *das Nennen der Götter,* and the poet's words are dependent on the *Winke der Götter*. As we saw above, only when the poet is receptive to the calling of the gods can he name them, in response to their calling. The power of the poet's words comes from the gods, but the poet does not simply receive the call and store it up for himself, like riches. The purpose of receiving and responding to the *Winke* of the gods is "*um sie weiter zu winken in sein Volk.*" [HWD,46] Thus the poet is also bound to the people.[20] Language is called *der Güter Gefährlichstes* because it opens the very possibility of Being being betrayed by beings. If the poet could remain with the gods, he would not need to fear such danger. Since he poetizes in language, however, he exposes himself to the danger that his words will be taken only in their common everyday sense, and that Being itself will thereby be lost from view.

The poet is thus thrown out into the "Between" (*das Zwischen*) of the gods and the people. And only in this "Between" is it decided "*who man is and where he settles his existence.*" [HWD,46] We recall that Heidegger, in *Being and Time*, defined Dasein as that being whose Being is still open, still at issue.[21] The place where that issue is decided is here called the "Between" of gods and men. The poet, as Dasein, gives voice to the tension which both unifies and separates gods and men by pointing to the temporality of the Between.

Finally, Heidegger returns to the title and beginning of his lecture. Hölderlin poetizes the essence of poetry, but not in the sense of describing a timeless concept. Hölderlin's texts call for a decision about poetry and thus about life. He does not simply supply one more indifferent concept (*"im Sinne eines zeitlos gültigen Begriffes"*), which may or may not be "adequate" to other available concepts of poetry. The essence of poetry belongs to a particular time, not in the sense of being "timely," but because through it a particular time finds its unique "word." This foundational character of poetry is the great theme of *"Wie wenn am Feiertage..."*. Whereas the "essence of poetry" taken as a static concept would lead to stultification and death, taken as a *Feiertag* it calls attention to the need for the right place and time for life to discover itself.

It must again be emphasized that man cannot choose his time. He is born into a time and must find a way to come to terms with it. Likewise with *Dichtung*: it too belongs to a particular "time" (or age). Heidegger shows Hölderlin's time to be a "needy" time, one marked by the absence of the gods. Yet, after all, the gods are not extinct (as in Nietzsche); their flight too belongs to temporality. What has fled will someday return. The gods have indeed fled, but even their absence is important for us because it is a fundamental aspect of our modern world. Hölderlin's poetry, according to Heidegger, recasts the world in terms of this double lack: that of the old gods who have gone and that of the "coming god" who has not yet arrived.

> Es ist die Zeit der entflohenen Götter und des kommenden Gottes. Das ist die dürftige Zeit, weil sie in einem gedoppelten Mangel und Nicht steht: im Nichtmehr der entflohenen Götter und im Nochnicht des Kommenden.
>
> [HWD,47]

The present age remains within the divine discourse as long as it "dwells poetically," i.e., as long as the essential word which brings man face to face with the nature of his existence continues to be spoken. This does not imply that Heidegger is simply waiting for a new god to come up so he can worship him. He says nothing about such a god, except that his mode of being for us is to be "coming" (i.e., to be "on the way" but not yet—always not yet—"here"). There is thus no criticism implicit in the term *dürftige Zeit*: it simply names our time as one of "want."

> Das Wesen der Dichtung, das Hölderlin stiftet, ist geschichtlich im höchsten Maße, weil es eine geschichtliche Zeit vorausnimmt. [HWD,47]

Hölderlin's poetry is not "historical" in the sense that it can be fitted into a pre-existing historical framework; e.g., the "early Industrial Revolution," or the "Age of Goethe." To see it that way would mean it was being seen through a medium which would deprive it of its very immediacy and value. Instead, Heidegger recognizes that Hölderlin lays down the essential parameters of the modern world (the *dürftige Zeit*) in his poetry. That is what makes his writing "historical."

Poetry's New Foundation

The purpose of this chapter, like that of the previous chapters, has been to seek a way of making sense of the temporality found in *"Wie wenn am Feiertage..."*. Unlike in poetry, the temporality of most expository writing is necessarily linear. A problem is stated, a methodology is chosen, data is collected, and results are examined. Only one sequence is permissable, if the exposition is to be clear. Of course, the proper sequence may be adapted to literary (non-discursive) ends, but such playfulness ultimately comes to rest on the presumptive linearity of the exposition. The temporality of *"Wie wenn am Feiertage...,"* however, is fragmented. It is essentially and ineluctably unclear when and where its temporality shifts from that of the *Feiertag* at the beginning to the cries out of the abyss at the end. Bringing two texts together with such different kinds of temporality as Heidegger's essay and Hölderlin's poem can lead to problems of incompatibility. One kind of temporality usually must give way before the other.

"Hölderlin und das Wesen der Dichtung" is an aberration from this tradition. It is not restricted by an allegiance to linearity nor by the boundaries of individual texts, nor even by the explicit meaning of particular passages. Rather, it sets out to show how Hölderlin is the "poet of the poet" by really listening to and thinking with his words. Its course is recursive and contradictory because it is unafraid of encountering the poetic text hermeneutically: as an open-ended,

reciprocal, and recursive approach-and-withdrawal of text and interpreter, in which contradiction and paradox do not end the investigation, but merely sharpen the focus. And yet it seems to come out with a reading of Hölderlin which does indeed work.

Heidegger sets out for the essence of Hölderlin's poetry. He does not take Hölderlin's texts as independent objects which can be fitted into a pre-existing schema; nor does he reduce them to historical artifacts, whose intrinsic interest might be overshadowed by their contextualization. Instead, he treats each of Hölderlin's "words" as fully alive and thus incommensurable with anything else. We must listen to them, think about them, respond to them. By allowing Hölderlin's words to affect his own thinking, Heidegger opens himself to a living encounter with Hölderlin's thought, and thus engages in a true dialogue with the poet.

This dialogue has unfolded in the form of a search by a thinker for the essence of poetry. We have followed the course of Heidegger's thought as it approached and withdrew from particular "texts" by Hölderlin. Frequently the texts contradicted each other; at times Heidegger seemed to take unconscionable liberties with the apparent, commonsense meaning of the words on the page. Each time, however, after such an arbitrary start, we saw him resist the urge to move on; instead, his thinking remained with those words and thus allowed something of their fundamental character to emerge. Undoubtedly Hölderlin would not have expressed it in such a fashion; but it was Heidegger himself who needed to understand Hölderlin's thought and that entailed remaining in dialogue with it until it could be assimilated to his own thinking.

Can we say after all this that Hölderlin does poetize the "essence of poetry," as Heidegger claims? Has he shown that? Poetry does indeed seem a game and yet it possesses an undeniably serious power over us. Poetry first brings "the gods" to language; i.e., it allows men to speak of the most important things, it gives them words with which to think. It thus brings Being closer. At the same time, however, it threatens to conceal Being behind words with a pretty façade to lull men back into their everyday, inauthentic modes of being. Poetry, according to Heidegger, is fundamental to man's understanding of language and thus of Being. When Hölderlin sets poetry on a new foundation (*"das Wesen der Dichtung neu stiftet"*), he brings Being into a new focus within which and from which a new historical mankind can reach understanding of the essential decisions facing it. As Robert Bernasconi puts it,

> The idea that Being is founded on the word of the poet is fundamental to all of Heidegger's thinking after this time [HWD], and the philosopher comes to be thought of in a similar light. Henceforth, Being is always thought of in respect of its coming to language.[22]

"Hölderlin und das Wesen der Dichtung" charts a new way of reading a poet. It does not focus on "texts" defined by literary scholarship but rather seeks to understand the poet himself, while at the same time eschewing biography. Where is the "essence of poetry" to be found? Most people and most literary scholars would look to the "classical" texts: Homer, Sophocles, Shakespeare, Goethe. They would point to works of undeniable scope and depth and beauty. Surely, in comparison with them, Hölderlin appears feeble and erratic. Although he wrote some beautiful verses, none of his works escapes a measure of deformity, insufficiency, obscurity, or (at times) disregard for formal harmony. None shows the scope of the *Iliad* or the *Divine Comedy*, none the *Weltfülle* of *Faust*.

Heidegger ignores Hölderlin's poems as such in this essay. Instead he looks for the "essence" of poetry, which is not assured by the "greatness" or formal perfection of a work and may indeed be totally incidental to it. Rather, as we have seen, the essence of poetry is bound up with temporality, though not as it is commonly understood. The usual questions about time are, "When?" and "How much?" Both are attempts at a quantification of time. Heidegger is not interested in such times. His question about time is, "What kind?" Temporality is central to his reading of Hölderlin because through this poet he is able to show (or remind) us how the poet "founds time anew." The essence of poetry is thus not found in pretty verses or formal perfection but in the "founding" of a revolutionary beginning in the life of a people. We shall learn more about this new beginning in the next chapter, which will follow Heidegger's reading of *"Wie wenn am Feiertage..."*.

As indicated at the beginning of this chapter, *"Hölderlin und das Wesen der Dichtung"* is valuable for our purposes because of the temporal focus guiding it. This is most clearly seen in the nonlinear, recursive path it follows to its goal. The "essence of poetry" brings man into contact with his existence in a fundamental way, by refusing to be pinned down and by establishing a new beginning for time.

> ...indem Hölderlin das Wesen der Dichtung neu stiftet, bestimmt er erst eine neue Zeit. Es ist die Zeit der entflohenen Götter und des kommenden Gottes. [HWD,47]

Such a new beginning does not merely set the clocks back. How many years (days, hours) is it until Armageddon? How can the time that separates us from the Garden of Eden be measured? As the pathbreaking poet Heidegger sees him to be, Hölderlin brings us and himself face to face with our age by speaking to us in a new way. The gods who are so often summoned in his poetry are inevitably the "gods who have fled." But Hölderlin does not mourn them or sentimentalize them. He returns to them looking for a new beginning, which requires a new god. A "new beginning" is a reorientation involving the whole "age" and all who dwell within it. It represents a rupture with what went before, a double "no:"

> Das ist die dürftige Zeit, weil sie in einem gedoppelten Mangel und Nicht steht: im Nichtmehr der entflohenen Götter und im Nochnicht des Kommenden. [HWD,47]

This is the great "Between" (*jenes Zwischen*) that was mentioned earlier, within which the poet must stand. It is similar to the *Ursprung* of the previous essay on art, in that it makes possible the artwork, while itself remaining impenetrable.

Did Heidegger actually reach the goal he had set for himself? He spoke at the beginning of the essay of his "intention" (*Absicht*) of showing the essence of poetry, and suggested a little later that the body of the essay was to be understood as an answer to the question of the status of Hölderlin's poetry: was it not

> ...das Anzeichen einer verirrten Selbstbespiegelung und zugleich das Eingeständnis des Mangels an Weltfülle? Über den Dichter dichten, ist das nicht ratlose Übersteigerung, etwas Spätes und ein Ende? [HWD,34]

These words are reminiscent of early criticism of Nietzsche, as well as of Hölderlin's own reception.[23] Early readers of Hölderlin were often all-too-ready to ascribe signs of madness to anything unconventional in his verse. We can easily discern Heidegger's inclination to irony here, as he recycles tired clichés about Hölderlin. How can a great poet be so wrapped up in self-reflection? Isn't that a sign of personal or artistic insufficiency? World-class poets are supposed to write about the world in all its richness, not about the craft of poetry. Once more at the essay's end Heidegger emphatically returns to the same words and asks the same question, but clearly with the expectation of a different judgment.

> Werden wir jetzt noch meinen, Hölderlin sei verstrickt in eine leere und übersteigerte Selbstbespiegelung aus dem Mangel an Weltfülle? Oder erkennen wir, daß dieser Dichter in den Grund und in die Mitte des Seins dichterisch hinausdenkt aus einem Übermaß des Andrangs? [HWD,47]

He does not explicitly affirm or deny either judgment; he simply leaves it up to the reader to choose which view of Hölderlin has more truth to it. Once again, though, it is clear where his sympathies lie. The superficial view of Hölderlin as self-absorbed and lacking a solid relation to the world has much on its side. Surely Hölderlin was especially introspective and labile in his relations with others. But it is hard to avoid agreeing with Heidegger that this poet speaks to us in a very fundamental way.

Far from being an "end," Hölderlin's words have been seen to be a new beginning. Like all great poets, Hölderlin redefines poetry by writing it and thereby challenges us to measure up to it. Heidegger has heard the challenge of Hölderlin's words and that is already a powerful answer. Nor does he fail his texts through a facile submission to traditional standards of literary propriety. The measure of Hölderlin's words can be taken by their effect on us: do we also choose to stand *"im Machtbereich der Dichtung"*?

Since we have found the "essence of poetry" to be temporal and thus non-conceptual, it seems the only adequate way to demonstrate an understanding of Heidegger's thought is to put it into practice. For the moment, however, I do not intend to go directly back to the text of *"Wie wenn am Feiertage...,"* although several areas of relationship readily suggest themselves. It is still too early, we are still too close to our original point of departure. Instead, I propose to read Heidegger's own, slightly later essay, *"Hölderlins Hymne 'Wie wenn am Feiertage...,' "* in the light of what *"Hölderlin und das Wesen der Dichtung"* has said about Hölderlin and poetry. Hölderlin's text will step a little closer: it will emerge as the focus of both of Heidegger's texts, as well as of our own. Reading them together with Chapter Two of this study should provide the impetus for another return to "the poem itself." As Heidegger wrote, with characteristically understated irony, in the Preface to the second edition of his *Erläuterungen*:

> Der letzte, aber auch der schwerste Schritt jeder Auslegung besteht darin, mit ihren Erläuterungen vor dem reinen Dastehen des Gedichtes zu verschwinden. Das dann im eigenen Gesetz stehende Gedicht bringt selbst unmittelbar ein Licht in die anderen Gedichte. Daher meinen wir beim wiederholenden Lesen, wir hätten schon immer so verstanden. Es ist gut, wenn wir das meinen. [EH,8]

Chapter Five

"HÖLDERLINS HYMNE *'Wie wenn am Feiertage...'*":

HEIDEGGER'S PROCESS OF READING HÖLDERLIN

Approaching the Essay

THE DECISION FACED IN *"Hölderlin und das Wesen der Dichtung"* concerned the nature of poetry and the power we allow it to exercise over us. Ultimately, it decided *"who man is and where he settles his existence."* [HWD, 47] Heidegger's reading of *"Wie wenn am Feiertage..."* [HH[1]] also confronts the reader with a decision, against which Hölderlin's poem is to be read. Here, too, the decision is not arbitrary or accidental: it is said to determine the form and character of man's existence on earth. Recognizing the necessity of a decision, however, does not imply that all that is needed is to "make" the decision. Having the "will to decide" is not enough: man's fate is not totally in his own hands, as Sartre would have it.[2] Heidegger is not evangelizing; he does not seek to confront an audience of sinners with the need for repentance. From his phenomenological-hermeneutic perspective, the inevitability of the decision *belongs* to man as part of his way of being. In *Sein und Zeit*, for instance, he writes, *"Es hat sich schon immer irgendwie* entschieden, *in welcher Weise Dasein je meines ist."*[3] Dasein is always "mine" in one way or another, which Heidegger goes on to call "authenticity" (*Eigentlichkeit*) or "in-authenticity" (*Uneigentlichkeit*). The point he seems to be making in this passage is that Dasein's Being-as-possibility allows it to *choose* how it will be—i.e., as itself or not.

Heidegger speaks in the opening words of *"HÖLDERLINS HYMNE"* of a decision of profound importance, yet does not specify what is at stake or what outcome he expects. He thereby risks arousing the impatience of many readers anxious to find out the "issue" so they can "make" a decision and move on to something else. But decisions of this magnitude cannot be made by individuals; I cannot "decide" the future of human history, for example. Rather, the decision should call attention to the background understanding which frames our reading of Hölderlin's poem.

Heidegger is calling attention, in my opinion, to the hermeneutic situation facing anyone who would read *"Wie wenn am Feiertage..."* in our age. "Making" decisions of this magnitude is not the issue; what is essential is rather first of all to *recognize* that a decision has already been made or is still "outstanding." Such recognition belongs to the phenomenological-hermeneutic method Heidegger employs. In the hermeneutic situation, the very act of understanding helps constitute what it understands. *"What is decisive is not to get out of the [hermeneutic] circle but to come into it in the right way."*[4] The method chosen is *"decisive"* (*das Entscheidende*) in that it first enables or excludes the possibility of discovering that which is being sought.

Since his investigation of *"Wie wenn am Feiertage..."* is to be hermeneutic, Heidegger's first question must be methodological: Where can access to the poem be found? As we have seen, Heidegger usually tries to begin with the obvious and everyday aspect of a problem. Thus, our question must be turned around and we must ask where the poem has *already* been found. If one is to approach *"Wie wenn am Feiertage..."* in a way appropriate to the way it is, one must allow oneself to be guided by it. As with all journeys, this one too begins with where one "is" or "is found" (*sich befindet*): caught up in everyday affairs, relying on common knowledge to attain generally recognizable aims. Such knowledge, however, hides as much as it reveals. Common knowledge takes Hölderlin's poem to be just another "thing," suitable, for example, for packing in soldiers' knapsacks.[5] But *"Wie wenn am Feiertage..."* will never be heard that way, according to Heidegger. It is not a tool of the discourse by which man subjects and rules the earth, turning the earth itself into just another resource for his projects. One cannot "order up" more *"Wie wenn am Feiertage...,"* like bullets or textbooks. By casting the decision in the words he does, Heidegger invites us to think about the very character of our language and of our being in the world. Far from being obscure or evasive, Heidegger refuses to rest content with the fleeting and superficial, or to join in propagating more mindless domination and destruction (*"das künftige Gepräge der unbedingt gewordenen Herrschaft des Menschen"*) by proposing a specific, apparently solid and reliable decision. The most dangerous kind of obscurity frequently lurks behind an unexamined certitude. To insert the decision about the "future shape" of mankind into a discourse of metaphysics would be to lose the ability to make the decision at all.[6] The decision called for by the technological age cannot itself be made through technology.

Chapter Two of this study focused on temporality in a poetic text. Temporality was thought to be something found within the text taken as a linguistic entity. What Heidegger has shown in the first two essays is how temporality influences even our initial access to the poem. *"Der Ursprung des Kunstwerkes"* began by questioning the familiar framework of form and content which takes the artwork as a "thing" merely standing present in our environment. Following his phenomenological hermeneutic, which lets *"that which shows itself to be seen from itself in the very way in which it shows itself from itself,"*[7] he proceeded to uncover the temporal aspect of the artwork's "work," in which it *"sets the truth to work."* Since the nature of truth (as *Unverborgenheit*) was found to be most purely preserved in *Dichtung*, the linguistic work of art, a path opened up which led to Hölderlin and which now leads back to the very poem with which we began.

"Der Ursprung des Kunstwerkes" and *"Hölderlin und das Wesen der Dichtung"* showed how Heidegger understands the hermeneutic process as recursive between question and answer, beginning and end, initial expectation and actual results. Both *"Hölderlin und das Wesen der Dichtung"* and *"HÖLDERLINS HYMNE"* begin with a text (a poem, letter, excerpt, etc.) in its everyday appearance. In *"Hölderlin und das Wesen der Dichtung"* the "text" was ultimately Hölderlin's entire *oeuvre*

and many readers had taken it to show him as unbalanced and narcissistic (*"das Anzeichen einer verirrten Selbstbespiegelung und zugleich das Eingeständnis des Mangels an Weltfülle"*[8]). In fact, however, through Heidegger's new reading of certain passages, a poet of fundamental insight and importance was uncovered. The present essay, on the other hand, restricts itself to one poem, which it comes upon within an historical transmission. The initial observations of each essay are almost absurdly simple-minded and clichéd. They obviously cannot be meant quite seriously as the author's own opinions. They are useful, though, as starting places for thought. The path of thinking must begin with the matter to be thought in its most common way of meeting us, but it must finally return to consider the very decision which instigated its journey.

This chapter will first attempt to follow the *process* at work in Heidegger's reading of *"Wie wenn am Feiertage..."* through a cursory structural overview. We have already followed a similar track in *"Der Ursprung des Kunstwerkes"* and *"Hölderlin und das Wesen der Dichtung."* The overview will show the introductory and concluding lines of the essay pointing to each other, and thus framing the interpretation of the poem, through the evocation of a matter awaiting "decision." Although little discussed in Heidegger studies, the concept of "decision" (*die Entscheidung*) is in fact crucial to understanding his approach to literature and especially to Hölderlin. If its significance is missed, Heidegger's thought appears not only obscure but even evasive. His vision of a decision facing modern man combats the accusation that his thought has no relation to praxis, without falling into the reductionist trap of judging the artwork through political or materialistic preconceptions. It does so by allowing those very terms to address the text and then working out the problems they encounter.

First Reading and Structural Overview of *"HÖLDERLINS HYMNE"*

For our purposes, *"HÖLDERLINS HYMNE 'Wie wenn am Feiertage...'"* can be divided into four sections: the printed text of the poem, an introduction to the essay, the reading itself, and the conclusion. (Heidegger himself did not explicitly mark divisions within the essay in any way.) Further distinctions could be made, but for now I wish to focus on only the major places of transition.

The poetic text printed at the head of the essay differs from that used in the rest of this study. It appeared in Hellingrath's edition of 1910 and is taken up without comment by Heidegger. In view of the state of the manuscript, much could be said about the choice of this text (and Heidegger's refusal to abandon it after Beißner's *Stuttgarter Ausgabe* began to appear, starting in 1940). Heidegger merely assures the reader that he has once again checked the text of the poem against the "sketches," presumably finding it acceptable (except for two places; see below).[9] The most significant differences between Hellingrath/Heidegger's edition and Beißner's occur at the poem's conclusion, where Beißner has "reconstructed" portions of St.VIII and St.IX, but which Heidegger largely ignores.

The first two paragraphs of *"HÖLDERLINS HYMNE"* introduce the poem by providing its *Entstehungs- und Rezeptionsgeschichte* in a few words. Heidegger describes the general poetic structure of the text, its lack of title, the number of strophes, verses, etc. Before proceeding with his interpretation, though, he calls attention to the need for a "decision" (*Entscheidung*) about the future of man. After mentioning it, he moves on without comment.

Thereafter and for most of the essay, Heidegger discusses each strophe in a linear sequence, as if simply reading through the text. When he comes to the eighth (and for him the last) strophe, he interrupts the linear progress of the interpretation to reread St.III,1–2 (*"Jezt aber tagts! ... das Heilige sei mein Wort"*) and to claim that *it* (and not the fragmentary manuscript ending) is the true conclusion of the poem, where it says what it has to say "authentically" (*eigentlich*) and therefore "to perfection" (*in der Vollendung*).

He returns to the same verses a third time to discover a path from the authentic perfection of the poem to us, NOW. That path concludes the essay by confronting us once again with a decision about man. The decision brings the poet at last into earshot of the contemporary audience.

The call for a "decision" at the essay's beginning and end serves not only to unify it or round it off, but to provoke renewed questioning of the relation of *"Wie wenn am Feiertage..."* to its "time." In both cases "Hölderlin's poem" or "Hölderlin's word" is read against an historical background in which a decision is awaited. The two passages of

"HÖLDERLINS HYMNE," although similar, are far from identical. The structural parallelism between the introduction and the conclusion has tipped us off to the importance of a "decision." We must read the essay a second time, paying more careful attention to the structural relations we discovered.

Second Reading

Heidegger's Introduction

Heidegger's readings of Hölderlin are often accused of losing sight of the (literal) text or of distorting it. What exactly is his method of reading Hölderlin? How does he even find Hölderlin's text? *"Wie wenn am Feiertage..."* seems to stand at the head of *"HÖLDERLINS HYMNE,"* its guarantor and source. But a text is never simply "found," as though it had an independent existence apart from a particular reading. A text speaks to us from somewhere, it requires mediation. Heidegger finds Hölderlin's text initially in the midst of its historical mediation.

He stresses the difficulties this seemingly self-sufficient text has faced in reaching its audience. As Heidegger points out, the poem, although written over a century ago, has still (up to 1939) not been adequately interpreted. In saying this, he ignores significant work by Lachmann, Beißner, Michel, and others.[10] Presumably their work was more editorial than interpretive. The mere passage of time is clearly not enough: the poem has required the intervention of a particular person, Norbert von Hellingrath, to give it a form (*Gestalt*) and to publish it. The poem apparently existed already, in a way, but was "formless and void:" even though the words were in existence and available (*vorhanden*) to anyone caring to read them in the manuscript, the very form of their self-presentation denied access to all but a few specialists. Heidegger casts Hellingrath in the "godlike" role of creating form out of chaos, thereby re-performing the original act of waiting/seeing/saying with which the poetic voice had characterized itself (*"Ich* h a r r t *und* s a h *es kommen,/ Und was ich sah, das Heilige sei* m e i n W o r t*"* [St.III, 1–3]).

Even Hellingrath's publication of the poetic text was apparently insufficient to ensure it a sympathetic hearing. In order to be truly heard, according to Heidegger, the poem needs more than simple linguistic existence, more than editorial clarification, which would only suffice to identify it as one example of a class of apparently interchangeable objects (i.e., "poetry"). Philological study remains deaf to what is truly important and unique in the poem. According to Heidegger, it requires and continues to lack interpretation (*Deutung*). At this juncture, *"HÖLDERLINS HYMNE"* offers its own "interpretation" (*Versuch einer Auslegung*) as a response to the poem's yet-unheard word.

What has impeded the poem from being heard for so long? History has not been kind to it. In a fascinating and complex passage, Heidegger suggests the historical context within which and in spite of which he reads *"Wie wenn am Feiertage..."*:

> Während dieser Jahrzehnte hat der offene Aufruhr der neuzeitlichen Weltgeschichte begonnen. Ihr Gang erzwingt die Entscheidung über das künftige Gepräge der unbedingt gewordenen Herrschaft des Menschen, der den Erdball im ganzen sich unterwirft. Hölderlins Gedicht aber harrt noch der Deutung. [HH,51]

While continuing to speak in terms of "years," "decades," and "history," Heidegger's discourse in this passage is beginning to shift from a materialist to a more "ontological" description. The words he uses here, which seem to refer to well-known current events or concerns of Heidegger's audience in 1939, actually evoke in rather dramatic fashion a fundamentally different view of history. It could be said that Heidegger indeed refers here to his own "time," but that time is precisely when "*the* decision *about the future form of man's unlimited mastery and subjugation of the globe*" (emphasis added) has been forced on all those living in it. The question is not *whether* to subjugate the earth, but *what form* the mastery of the earth will be given in the future. Man here assumes the dimensions of a Promethean artist.

The present age is thus not just the latest in a series of empty temporal units: the "*twentieth* century," for example. Nor can the distance between the poem and the present be expressed in a number of years. The context up to this point in the essay has been the language and methodology of the philologist, objective and corroborable. Heidegger has listed the poem's date of composition, the interval until its publication, and the name of its publisher. He has indicated the nature of its reception and the likely reasons for it. The tools of the philologist, however, leave the poem unheard. It cannot compete for the attention of "the Germans" (i.e., of its audience) with the claims of the technologized world, which have now reached a deafening volume. The poet's voice is too feeble to pierce the sound barrier erected by industry, entertainment, armed conflict, etc., which daily clamor for people's attention. His words are too strange, too quiet, they make readers "lose" too much time. After all, it is well known that the German people (among others) presently (1939) faces an historical crisis. What is needed, people say, is action, not reflection—and certainly not poetry.

The urgency with which this decision addresses us, however, bespeaks its deceptiveness. What exactly is to be decided?

1. The FUTURE, that which is "coming."

2. Mankind's MASTERY (*Herrschaft des Menschen*), a phrase in which the ambiguity of domination is scarcely concealed: does man rule, or is he ruled; and if so, by whom?

3. Mankind's mastery over the EARTH (*den Erdball*), taken as a lump of material, a resource.

4. The UNBOUNDED (*unbedingt*) quality of that mastery, and

5. The CHARACTER (*Gepräge*) of that mastery by/of man, which forms the subject of the decision.

It seems as though Heidegger were describing the present from within a metaphysically contaminated discourse, namely history taken as the record of Western civilization and its technological advances. His words, though rather obscure, could lead some to conclude that Heidegger was welcoming the rise and accelerating pace of industrialization, including possible experiments with human genetics, in order that a *Herrenvolk* might emerge which would control *absolutely* the earth and every "thing" on or in it. Even his vocabulary is strikingly harsh: *Aufruhr, erzwingt, Herrschaft, sich unterwirft*. In quiet contrast, the last sentence of the paragraph lets the hollowness behind such harsh claims come through: *"Hölderlins Gedicht aber harrt noch der*

Deutung." This sentence might seem a rather pale contrast with the drama of the preceding pair, just another reworking of a trope familiar from devotional literature: the "still, small voice"[11] which guides one through tumultuous times. Poetry has often been seen as a refuge from the difficulties of living in the "real" world. What possible relationship could exist between Hölderlin's poem and contemporary world history?

In fact, a curious dialogue about time has been taking place in the course of the first paragraph. Two ways of talking about time have been juxtaposed: the *chronological*, resting on assumptions of the physical sciences and discursive reason; and what, for lack of a better name, must be called the *ontological*, which views time in terms appropriate to its essential nature; i.e., from time itself.

Chronological time takes a particular event as an absolute beginning and restricts all talk about time to questions of priority or sequence. We have seen how, from the standpoint of world history—i.e., of time considered *chronologically*—this poem is a pretty sorry thing; it was insignificant at its conception, never achieving "final," or publishable form; its textual condition is a shambles; and, even after being edited and published, it has elicted (to 1939) little significant response or understanding among readers. It seems to face a future of continued obscurity in some archive, as little known as in its author's lifetime. Meanwhile, historical forces have grown ever more powerful and inescapable. The time is approaching for a final confrontation, *the* decision about the future of mankind, in which a few lines of poetry can have nothing to say. In historical time the poem never was, nor can it ever be a significant factor. It seems quite remote from our most pressing concerns.

Ontologically, however, the poem has been approaching us since its inception, and now the time for its true audience seems to have arrived. As Heidegger narrates it, the poem has acquired existence (*entstand*) and form (*Gestalt*); it has passed through history measured as ages of man (*Menschenalter*), finally arriving here before us today in this lecture, as this specific text. It is important to realize that Heidegger's own narration does not simply report or announce the arrival of the text, but itself independently reconstitutes that arrival. Through this essay, *"Wie wenn am Feiertage..."* is brought before us, though not as a "thing" to be examined for its qualities or its condition, like a vase. Rather, at the introduction's end the poem has first been revealed as something outstanding, a question yet to be answered.

Thus Heidegger has been engaging in a double discourse in the first few paragraphs of the essay. While it appeared at first that the poem had been found in the midst of a unidirectional historical process, he has discovered a different kind of temporality at work. Time that allows itself to be measured (chronological time) exists together with "special" times (ontologically: the *kairós*), which elude all measurement. The "right" time for doing something is unpredictable; it may come when we least expect it, and it may fail to come when all signs point to its arrival. One can speak of a quantity of time ("20 years"), but no quantity can "add up" to the time for success or failure. The *kairós* comes to me as a gift; I cannot force, compel, or even "persuade" it to come. Heidegger implies the *kairós* for reading *"Wie wenn am Feiertage..."* has arrived.

The purpose of the introductory paragraphs of *"HÖLDERLINS HYMNE"* has been to "find" or "locate" Hölderlin's poem. It seems only natural to find it in the present, as an historical artifact. The poem is simply "there" as stable, knowable; as something which has always been there, and always in just this form. Heidegger has shown in a few words how deceptive such an impression is. In fact, according to him, the poem's existence has been and continues to be marked by its inaccessibility; it has even yet not been heard. *"Hölderlins Gedicht aber harrt noch der Deutung."* Hölderlin *himself* may not have heard what his own poem says, as in his use of the word *Natur* [HH, 57f]. The artist, after all, does not have a hegemony over the work of art; the fact of having "made" it (*poiesis*) does not entail having comprehended it in its essence (*ein wesentliches Verständnis*). Rather, each finds its origin in the other, the artist as well as the artwork (cf. *"Der Ursprung des Kunstwerkes,"* 1).

Thus, it is quite understandable that Hölderlin's poem has not yet been heard, since even the poet has failed to fully hear it. Nevertheless, as Heidegger claims elsewhere, Hölderlin's verses are "destined" for us, only we can read them, they are uniquely our own. *"Hölderlins Dichtung ist für uns ein Schicksal. Es wartet darauf, daß die Sterblichen ihm entsprechen."*[12] This destiny does not mechanically enslave man, or predestine his every action. It does not rob man of his ability to *decide*. Indeed, man's destiny is dependent on his temporality: only in that time when *"die Sterblichen ihm entsprechen"* does his destiny step forward. A decision seems to be called for—"called" not in the sense of requirement or duty, but as a "calling" which awaits a response (*eine Entsprechung*).

First Return to St.III
[HH, 72–5]

A prime example of Heidegger's seeming arbitrariness becomes apparent through consideration of the structure of *"HÖLDERLINS HYMNE"* as a whole. As remarked above, he reads through the poem linearly, after establishing a context for us "modern" (*neuzeitlichen*) readers. After reading up to the eighth strophe, however, he discards this relatively straightforward approach and returns to St.III,1–2, which says the "essence" of the poem. This enables him to read St.VIII,1–3 (where his text breaks off) as monitory but ultimately hopeful, in spite of textual evidence for the inclusion of several additional lines (including parts of a final St.IX) which would leave an opposite impression. In the commonly-accepted version of the text (*Stuttgarter Ausgabe*, II/1, 118–20), the poem ends with a repeated cry of despair (*"Weh mir!"*), *not* with a re-affirmation of the poet's privileged relation to the immediate (to Being).[13]

The rhetorical move by which Heidegger accomplishes this new reading of the poem seems quite strange. After correcting Hellingrath and Zinkernagel, the editors of his version of the text, for sloppy philology, he himself then proceeds to offend against philology by inserting a period where the text shows none, with the sole (and only implicit) justification, that the "thinking" of the poem demands it.

> Mit Vers 63 [St.VII,1] beginnt ein Denken, das in das Sagen des Heiligen zurückkehrt und die Vollendung des Gedichtes einleitet. *Deshalb* wurde im vorliegenden Text an das Ende von Vers 62, das bei Hölderlin ohne Satzzeichen geblieben, ein Punkt gesetzt.
>
> [HH,72; emphasis added]

The inconspicuous period (*Punkt*) points clearly to the priority of the poem's thought to its philological presence. As far as Heidegger is concerned, if the text's thought requires a change in its appearance, so be it. He succeeds in this move only to the extent that the reader is willing to enter into his interpretation of the poem's "thought."

What does the move back to St.III accomplish? The linear interpretation was about to reach its "end." The significance of what it said would be heightened by its position *as* end, as the final lines of the poem. Read naïvely, the last lines cast a serious shadow on Heidegger's fundamentally positive reading of the poem. Heidegger, however, does not read naïvely; he listens not only to what the poem does say, but also to what it does not say. The second reading reads the "last" words as "inconclusive"; they are only an accident of history. That is, Hölderlin did not in fact complete the writing-out of the poem, but even if he had supplied the final lines, the poem would not have said "more" than it already does, at least in Heidegger's view.[14] The interpreter must return to the "source" of the poem, where the true "end" (*die Voll*end*ung*) speaks.

Heidegger thus commits another offense against standards of argumentation. He has already: 1) focused solely on the text that best supports his own preferences; 2) changed even that text (through addition of a period); and 3) simply *ignored* the state of the poetic text when it interfered with his own interpretation. A reading of St.VIII,1–3, such as in Chapter Two of this study, shows a chasm separating the poem's beginning and its end. *"Wie wenn am Feiertage..."* begins joyously and hopefully, but, with the introduction of Semele, turns dark and menacing (even leaving aside for now the remaining verses not discussed in *"HÖLDERLINS HYMNE"*). Heidegger is not simply being arbitrary; he has maintained throughout the essay that philology and rationalism cannot reach the poem. By not allowing philology to dictate to him, Heidegger reminds the reader that he is doing more than writing a commentary on the poem; he is after its "essence" (*"die Vollendung des Gedichtes"*). And that essence is to be found in St.III:

> Denn was diesem Gedicht eigentlich und sonach in der Vollendung zu sagen aufgegeben ist, das sagt es selbst in der alles tragenden dritten Strophe...
>
> [HH, 72]

Heidegger does eventually read the eighth strophe; it is not suppressed, only delayed. Before reading it, though, he wants to remind us that we must approach it hermeneutically. The interpretation has had to read the entire poem up to this "point" (*Punkt*) to realize the importance of those lines in the third strophe. Because of their "centrality" they must be brought back to the "center" of the poem, where it "*says what it has been charged with saying authentically and therefore completely.*" Merely reaching the "end" of the words on the page does not suffice to "end" the interpretation. That is, for the poem to truly end it must return to its origin, source, or essence (*Wesen*)—which Heidegger finds in *"das Sagen des Heiligen."* [HH,72]

Second Return to St.III
[HH,72–75]

In St.VII the poet speaks of the dangers arising from contact with the divine. Heidegger does not deny the existence of danger in such contact, but sees its direction reversed: rather than the gods threatening the poet, it is the poet (and his "Word") who threatens not just the gods, but the "Holy" itself. Obviously, no mortal poet can "threaten" the gods; but he can do something worse. He can (intentionally or not) betray the very nature of the Holy (and thus of the gods) by depriving it of its immediacy, by making it into a "thing" (e.g., a "word"), something mediated, and thus the opposite of its essential nature.[15] The danger of the word can be seen in a national anthem, for example. It threatens to replace something valuable though intangible (patriotism, group solidarity) with concrete words or images, which may be manipulated in ways contrary to the "essence" of their referent. Thus, instead of bringing the people together and leading them back to their origin in their particular "earth," it might be debased into a tool for political ends (e.g., anti-Semitism). In this way, even words spoken in response to the Holy can constitute a threat to the Holy.

"Das Heilige ist die einstige Innigkeit, ist 'das ewige Herz'." [HH,73] The Holy is here seen in two ways: 1) as *Innigkeit* (literally "innerness," but also "heartfeltness"), it reminds us that human beings are never simply isolated things "out there," but rather always exist "in" a world to which they belong. Dasein's capacity for "Being-in" creates the possibility of its understanding its world. Hölderlin's word for it is *"das Herz."* 2) The Holy is also seen as *"das Einstige," "das Anfängliche"*—that which exists as absolute anteriority and "originality." The Holy, as the source of our sense of "belonging-to" something, cannot be harmed

by poets or gods. The ashes of our ancestors will always remain "holy," in part because no power on earth can *harm* them; they will always be "whole." After learning to see the Holy in this way, the essay proceeds to read the poem's "final" words:

> Des Vaters Stral, der reine versengt es nicht,
> Und tieferschüttert, eines Gottes Leiden
> Mitleidend, bleibt das ewige Herz doch fest. [St.VIII,1–3][16]

By focusing on these words and leaving out the fragmentary last lines of the text, Heidegger is able to transform the character of the poem from problematic and somber to resolute and hopeful. This version of *"Wie wenn am Feiertage..."* seems to say that the poet does succeed in mediating the divine, whereas the commonly-used Beißner edition leaves little room for such a view. As we have seen, however, Heidegger does not take these as the truly "final" words, and he does not interpret them in terms of St. VIII, but of St.III. The danger is not to the poet but to the Holy.

Heidegger admits the poem lacks an ending, but asserts that this lack arises from the superabundance of its origin (*Anfang*). He thus finds himself called once again to the "essence" of the poem, to "what it has to say," to its "Word:"

> Jezt aber tagts! Ich harrt und sah es kommen,
> Und was ich sah, das Heilige sei mein Wort.
> [St.III,1–2]

These lines, as the poem's "source," cannot be exhausted or find an "end." Having listened to what the poet says about the Holy and the poetic "Word," Heidegger has still not finished. The Word declares itself to us NOW (*jezt*). But the temporality of this NOW is still obscure.[17]

> "Jezt"—wann ist dieses "Jezt"? Ist das der Zeitpunkt um 1800, da dies Gedicht entstand? Das "Jezt" nennt ja doch eindeutig den Zeitpunkt, in dem Hölderlin selbst sagt: "Jezt aber tagts!" Gewiß nennt das "Jezt" Hölderlins Zeit und keine andere. Aber Hölderlins Zeit ist ja einzig die durch sein Wort angestimmte Zeit. Hölderlins Zeit ist allerdings im

> strengen Sinne seine Zeit. Aber diese seine Zeit ist gerade nicht das zu dieser Zeit nur gleichzeitige und übliche Zeitgenössische. [HH, 75–6]

Thus, the date of conception (in which "now" stands for "1800") is not adequate for describing the NOW of the poem. It is a mere *"Zeit-Punkt."* Were it ever held to be adequate, it would translate the poem into an unapproachable remoteness, where it could have little relevance for a future reader. But the poet's "now" is just as far removed from "1800" as from "1939." Taking *Jezt* as an utterance of the poet "Hölderlin," Heidegger affirms that it is "his" time, but differentiates between that time and one which is merely simultaneous (*gleichzeitig*) with the historical Hölderlin.

The NOW of the poem opens the realm of historical decision by naming the Holy. History, conceived as an empirical science which describes events in terms of chronology (*Historie*), is inadequate to Hölderlin's poem. What is needed is another kind of history (*Geschichte*), in which the *kairós* may appear (or fail to appear), and which can only be approximated in English by a phrase like "at this *historic* moment." Insofar as *Historie* purposely avoids discriminating against any applicant to its roll-call, it overwhelms whatever has true *historical* significance. The decision for one or another historical direction comes more often than not in a small, quiet, unobtrusive moment, which may actually be overlooked or misunderstood by its contemporaries. As Heidegger explains,

> Dieses Historische ist aber nie die Geschichte selbst. Geschichte ist selten. Geschichte ist nur dann, wenn je das Wesen der Wahrheit anfänglich entschieden wird. [HH, 76]

This other, rarer, kind of history *decides the nature of truth*. Heidegger does not specify *how* the decision gets made, nor indeed can he. It does not happen every day; it cannot be found by collating all the headlines from a year's newspapers or by taking a poll of important personages. That is, such history is not a function of "great" men or women. Even so, when history (*Geschichte*) does occur it changes everything, irrevocably. The decision as to what will count as important and what will be irrelevant is the "naming" of the Holy, which should not be confused with the establishment of a religion. We should recall here what was said about history in *"Der Ursprung des Kunstwerkes:"*

> Geschichte meint hier nicht die Abfolge irgendwelcher und sei es noch so wichtiger Begebenheiten in der Zeit. Geschichte ist die Entrückung eines Volkes in sein Aufgegebenes als Einrückung in sein Mitgegebenes. [UK,63]

"History," in Heidegger's use of the term, always implies the original setting of parameters for a historical people. It gives a people its "task" (*sein Aufgegebenes*) as something which seems so natural as to be simply "there" (*sein Mitgegebenes*). At a certain time, people come to recognize certain obligations as unavoidable and necessary and may even be puzzled that they had not previously been concerned with them. How could people for so long "ignore" environmental issues, for example? Heidegger makes yet more explicit what he means by this decision when, a few lines later, he repeats the same formulation:

> Das Heilige *entscheidet anfänglich zuvor* über die Menschen und über die Götter, ob sie sind und wer sie sind und wie sie sind und wann sie sind.
> [HH,76; italics added]

The Holy itself decides: no longer does man appear as the autonomous subject of history, as at the outset of Heidegger's inquiry. He is no longer engaged in a struggle for global mastery. He has become only one side of a symbiosis with the gods, in which both are separated from immediate truth—gods as well as mortals. Like Zeus, they may each hold the scales of power in their hands, but the outcome gets decided *anfänglich zuvor* by a darker, more enigmatic power. The verb here, *entscheiden*, which I have translated as "decides," is not the more common reflexive (*sich entscheiden*), which usually expresses a subjective intentional state of the speaker. It is rather a transitive verb used by judges or referees to give clear and final form to their judgement in a *Zweifelsfall*, or case of doubt. This kind of decision decides what world will count as "true," regardless of how or by whom it gets made.

Nor is the string of dependent clauses (*"...ob... wer... wie... wann..."*) mere rhetorical flourish. Each clause brings out a different decision, probing the nature of the relationship between men and gods a little further. Most fundamentally, the Holy decides whether or not they exist at all (*ob sie sind*)—but not as an answer to a question like, "*Do men and gods* [i.e., do I] *exist?*" Such a question was faced, for example, by Descartes. Through reflection he concluded he had been able to achieve certainty about knowledge of his own existence. Heidegger is not attempting to answer such a question. Dasein's factical existence precedes any question about it. The decision which founds that existence, according to Heidegger, originates in the Holy.

Secondly, the Holy decides that men and gods exist "as" something, it gives them identities (*wer sie sind*). Man does not live in a world of faceless objects. He "identifies" or names particular individuals in his world and thereby allows them to matter to him (and to enter into his own "identity"). Man always exists in a world of concerns, even if they are not always thematized explicitly.[18]

Thirdly, the Holy decides "how" gods and men will matter to each other (*wie sie sind*)—whether they will live closely together, as they did in Abraham's or Jesus' times, perhaps; or whether they will live in the "*Nähe der entflohenen Götter: in dem Raum ihrer uns schonenden Flucht.*"[19] At different times a people may feel itself to be surrounded by its gods, even led by its gods in that people's own "historic" mission (as, for example, the Islamic fundamentalists). At other times, the gods may seem to have deserted them, to have absconded. These are two of the possible ways gods and men may live together.

Finally, the Holy decides that gods and men shall live in a historical time (*wann sie sind*). "When" does not point to a chronological time, as in the B.C./A.D. schema. Man exists, when he exists, within a time. This clause points to man's "thrownness" (*Geworfenheit*), such that man does not "create" or "choose" his time, but rather finds himself always already in a time. Moreover, the Holy decides "when" we live in the form of a question, or more precisely, a challenge. When do we exist? Do we exist now? Who are our neighbors in time, who counts as present for us and who as absent yet not forgotten? Of course, most of us do not consciously concern ourselves with "when" we live in such a manner. We are involved in our daily concerns, as we must be. Yet the background of those concerns necessarily involves an answer to the question "When?" Fundamentally, however, "when" refers to the nearness or remoteness of beings with which we are involved. Even so, the "when" which has been assigned to us by the Holy always remains available, should the right "time" (*kairós*) arrive. The Holy, in determining "when" man exists, determines the form history can take, if it should occur.

The Conclusion of "HÖLDERLINS HYMNE" [HH 75-77]

At the conclusion of the essay, Heidegger returns to the context of concerns with which he began. Hölderlin's poem, which had appeared to be insignificant and helpless when compared with the monumental historical processes (i.e., the rise of nationalism, the growth of technology, the capacity for "total" war, etc.) that had developed in the years since it was written, has been seen to speak in a radically "new" (*anfänglich*) way of man's being "on earth." The word of *das Heilige*, which is Hölderlin's word, recalls man to his history, decides in advance the parameters under which things like "nation," "technology," or "war" can have any meaning at all. This message is not hidden somewhere in the text or in its reception, it does not require "research" or "work" to be discovered; it only requires a certain *time*. Heidegger's sensitivity to the presence of temporality in language has led him to the kind of temporality at work in *"Wie wenn am Feiertage...,"* which is not one of dates or time-spans. Man's way of being in such a time is one of waiting (*harren*) for the "right" time, the *kairós*. Such a time cannot be compelled; rather, its arrival compels *us*.

To understand this better, we need only read again the opening words of the essay. We find there a description of a decision which was being called for by history:

> Während dieser Jahrzehnte hat der offene Aufruhr der neuzeitlichen Weltgeschichte begonnen. Ihr Gang erzwingt die Entscheidung über das künftige Gepräge der unbedingt gewordenen Herrschaft des Menschen, der den Erdball im ganzen sich unterwirft. Hölderlins Gedicht aber harrt noch der Deutung. [HH, 55]

In this initial formulation a decision is enforced or compelled (*erzwingt*) as to the shape of the future. It is posed by the growth and tumult of the modern world, which recognizes only "new" or "old" times, each identical with all others in its emptiness. "1914," after all, is equally as empty as "1910," when viewed merely as a measure of time. This "modern world history" is approaching a decisive turn: one way or another, the future of man's mastery over the subjugated earth will be determined. Such a view, in which man's mastery over his world is unaccountable and unrestrained (*unbedingt*) because it recognizes only force, is in fact a commonplace in our time, as it was in Heidegger's. Few would likely argue with its apparent content, although the words, when taken literally, are more suggestive than precise.

The significance of the state of affairs in which this decision will be made is put into a different perspective by the last sentence: *"Hölderlins Gedicht aber harrt noch der Deutung."* No amount of technological mastery of things will suffice to understand or subjugate this poem. The dissonance set up between Hölderlin's poem and the common understanding of "history" (*die neuzeitliche Weltgeschichte*) supplies the momentum of the essay. But does this commonsense understanding approach history in a way commensurate with it?

"HÖLDERLINS HYMNE" as a whole argues forcefully against such an interpretation. As if to underline this point, Heidegger returns in the last two paragraphs of the essay to speak again of the decision facing mankind:

> Hölderlins Wort sagt das Heilige und nennt so den einmaligen Zeit-Raum der anfänglichen Entscheidung für das Wesensgefüge der künftigen Geschichte der Götter und der Menschentümer. [HH, 77]

These words clearly respond to those at the essay's beginning. They return to the decision which set forth the historical field in which gods and men first came together. They also assert the continuing importance of that decision as long as there shall be "history." History, however, is no longer the inexorable force it first appeared to be. It is rather the story of gods and "humanities." The use of the plural form (*Menschentümer*) underscores the openness of such history. Neither men nor gods take up the center of this discourse; they are brought together by the historical decision.

How the decision gets made is not said. Rather, the fact of its being made supersedes the manner of its coming to pass. Through its ability to decide primordially the nature of historical truth, the poetic Word *"ist geschichtlich im höchsten Maße, weil es eine geschichtliche Zeit vorausnimmt.*" [HWD, 47] It does not look to the future as simply more of the present, but rather *founds time anew*, thereby enabling "the essential-coming-together" (*Wesensgefüge*) of gods and men, instead of the arbitrary mastery (*Herrschaft*) of man over the earth. It is a decision *for* (*für*) and not just *about* (*über*) the future. Access to the realm where this decision is to be made is granted by the poet, not by the irresistible advance of *"der neuzeitlichen Weltgeschichte."* Time and space, instead of closing us off from the poem, themselves receive identity and existence through the word of the poet. They exist as places for the coming-together of gods and humanities. It is only through the memory of such occasions of coming-together (Latin: *coetus*) that "chronological" time can be measured.

Third Reading: Hölderlin's Poem and the Decision

The issue in *"HÖLDERLINS HYMNE"* is to hear Hölderlin's poem as it truly is, to hear its "Word," as Heidegger calls it. In order to hear the poem rightly, it must first be found. The essay finds the text as though it were something familiar: a "poem" embedded in a certain process of creation and reception, and "fixed" on the page. This initial impression is quickly shown to cover up the level of discourse at which essential decisions are being made about man and the kind of world he will live in. We must *learn* to listen to poets like Hölderlin, Heidegger is saying; we must let them teach us how they want to be heard. The more patiently we remain with the poet's "Word," the less we are deceived by the appearance of his "words," and the less importance is attached to the text's historical status. We learn to hear the world speaking through the word of the poet, rather than trying to hear the poet through the words and apparatus of historians.

Heidegger has been accused of making Hölderlin a priest for a new disclosure of Being, for a new parousia.[20] Certainly the formulations of *"HÖLDERLINS HYMNE,"* if taken in their everyday sense, encourage that interpretation. For example, *"Das Heilige verschenkt das Wort und kommt selbst in dieses Wort."* [HH, 76] But this chapter has endeavored to show how important a reading of Heidegger's *process* of thinking is to understanding the words he uses. Heidegger's words are not universal currency. They cannot be removed from their context for exchange in other discourses. They do not "stand" for something.

When Gerald Bruns claims that *"Heidegger doesn't have a method"* in reading poetry,[21] he is emphasizing Heidegger's freedom from scientific philology with its demand for submission to the material circumstances surrounding the text, its genesis, and its reception. A few sentences further, though, Bruns makes the stronger case that *"[n]othing [Heidegger] says about a poem can be confirmed by analysis."* Here I believe he disregards the subtle but crucial place taken by "method" in Heidegger's readings. Heidegger is not being mysterious or evasive; and although he does indeed seek to "estrange" us (Bruns' term) from inappropriate methods of reading poetic texts, he also shows evidence of employing a recognizable approach. His reading of Hölderlin takes place within a hermeneutic process and only makes sense when understood from within that process. I have called this process "hermeneutic" because it tries to understand its own foundations as a result of its reading. It thus must take each insight, each supposedly definite result, as only temporary. Each thought circles about the central thought, which is contained within the beginning of the entire endeavor to understand. *"HÖLDERLINS HYMNE,"* viewed hermeneutically as process, hears the poem speaking from a historical "crisis."

Insofar as the present study shares this phenomenological-hermeneutic approach, it too must return to its beginning. Where is it to be found? Shall we simply read the text of *"Wie wenn am Feiertage..."* again? *Can* we do so? I think not. The poem as we found it in Chapter Two has been irretrievably lost. To find it now would necessitate forgetting what has been learned from Heidegger. Instead, our task is to read the text with Heidegger in mind.

By viewing history through this poem, Heidegger's interpretation of *"Wie wenn am Feiertage..."* reverses a powerful prejudice. *"Wie wenn am Feiertage..."* has heretofore always been approached through its historical record. That record, however, does not record what the poem truly says, it does not and cannot hear the poem's "Word." Obviously, history is written with the words, ideas, and images of the poets. The poets have given them to us and we have accepted them as our own, as though we could "have" them or "lose" them. Now things have gone so far that some of us believe we can pass judgment on poetry itself. Can the river drown its source? *"So vermag das dem Ursprung Entsprungene nichts gegen den Ursprung."* [HH,74]

Chapter Six

Conclusion: Temporality, the Poetic Text, and the Decision

As noted in the Introduction to this study, temporality in literary texts can be discussed in a variety of ways. Most often it appears as part of a larger critical apparatus. When used in that way, it does not usually present too many problems and seems clear enough. But when it is shifted to the center of the study, "temporality" becomes quite problematic, as Augustine already knew.[1] Each researcher seems to have a different concept of it. There is even some doubt that temporality can ever be adequately conceptualized. As a consequence, the investigation of temporality is cast back on itself, it is forced by the nature of its subject to re-examine its own preconceptions. Is there a single, unified concept which can be called "temporality," and which can be usefully made the subject of study? To ask the question thus is to realize its inadequacy and to make the first step towards clarifying it. In this way, temporality brings up the issue of process.

If we confine ourselves to literary texts, what kind of process do we find? There is the linear process of reading a text through from beginning to end. There is also the recursive process of returning to the text, no longer naïvely, and so able to see connections, both pro- and retentive (to speak with Husserl[2]). Perhaps more fundamentally, there are often references to historical tradition and to contemporary expectations; we find self-advertisements, which no text can escape. Although these references are more fundamental than the immanent textual process, because they first make possible our access to the text, they often only emerge subsequently to a reading. That is, the historical embededness of a text is seldom so apparent or insistent that it is processed along with a first reading of the text. Upon reading a line of Shakespeare, for example, one may be struck by its strangeness or "quaintness" of expression, which initially may impede or obstruct understanding, but only subsequently does its historicality come to the fore explicitly. Its strangeness has now acquired a name and an identity. The discovery that a text is historical, i.e., not vaguely contemporaneous with the reader but belonging to another historical time, is already a second stage in an attempt to make sense of it. Once an appropriate historical framework can be devised, the text may no longer seem so difficult to understand.

Reading a text for its internal structure and for its historical status are both ways of accounting for temporality through process. But temporality also enters a text as subject or theme. The awareness of "time passing," of the *volatilis aetas*,[3] is one of the oldest themes of world literature. Texts as widely separated in time and space as the *Bhagavad Gita* (c. 2000 B.C.E.) and the *I Ching* (c. 1250 B.C.E.) show a central preoccupation with this idea. Indeed, one way of looking at texts as such would be as an attempt to overcome or negate temporality. This aspect of literary temporality is invisible to a structural investigation, because it escapes the referential nexus which such an investigation employs. Temporality as a concept must be read out of words and actions; seen structurally, "time" is just a word like any other. As a topic, however, its importance arises from what is said about it, not from how it is said.

"Wie wenn am Feiertage..." presents difficulties for both approaches to temporality. Its structure exhibits considerable complexity and ambiguity, even apart from the fragmentary state of the manuscript. Its historical status as a modern attempt at a Pindaric hymn also lends complexity to understanding its relationship to temporality. This study has taken the poem initially as it is most readily found, in the standard version based on the *Große Stuttgarter Ausgabe*. That is, textual and historical problems were deliberately left aside. That version (actually

reconstruction) was printed at the start of this study. Its value was that it gave a coherent form to Hölderlin's words; its drawback was that it covered up the actual state of those words (in the manuscript), which is anything but coherent. As insufficient as the Beißner edition may be, it—or somthing like it—must clearly prepare the entrance to any serious approach to the poem. We must recall that the goal of the study was to observe temporality at work in poetic language. In the course of that examination, much was learned about the temporal possibilities of poetic language, but the poem itself seemed to recede and to deny our attempts to "grasp" it. It seemed clear that our own methodology needed review. We found apparent divisions in the text itself, self-contradictions which cast some doubt on the very quality of the poem *qua* artwork. Turning to Heidegger for some guidance in this matter, we were forced to reconsider the nature of the artwork that *"Wie wenn am Feiertage..."* is. This did not mean we had misread it (as a poem), or that by ignoring its historical background we had misinterpreted its words or the purpose of its composition. Rather, the problem was that our approach to *"Wie wenn am Feiertage..."* had understood it as a kind of "thing"—beautiful perhaps, and finely made, but ultimately something *zuhanden*: present-to-hand and available for quoting on suitable occasions or for solitary delight. This was exactly as if we had sought the label of Van Gogh's shoes.[4] The process at work was one of importing or incorporating *"Wie wenn am Feiertage..."* into "my" world. It was adapted to "my" views, and if it was a particularly good example of an artwork, it might add something to "my" experience. Ultimately, though, "I" remained unchanged.

If, on the other hand, this "I" steps forward into the world of *"Wie wenn am Feiertage...,"* if "I" allow the poem to speak to me and consider it important enough to warrant an answer (if I "respond" to it—*ihm entsprechen*), without expecting that my answer will exhaust its question or its "saying," then the poem will have been read as an "origin:" *"eine ausgezeichnete Weise, wie die Wahrheit seiend, d.h. geschichtlich wird."*[5] Have we yet entered the realm of the poem as an origin? This chapter must show how that can happen.

Heidegger's language and method were idiosyncratic enough to warrant considering two of his more general essays on art and on Hölderlin before looking at his reading of *"Wie wenn am Feiertage...."* It is tempting, but misleading, to call them "theoretical" essays, since they in no way oppose themselves to applied or "practical" works.

Instead, we may call them more appropriately "pathbreaking" (*wegweisend*). They accustom us to Heidegger's hermeneutics.

"Der Ursprung des Kunstwerkes" showed the inadequacy of traditional concepts used to understand artworks, particularly approaches relying on a distinction between "form" and "content." They were seen to rest on metaphysical conceptions of subject and object which reduced the artwork to a kind of equipment (*Zeug*). What was needed, according to *"Der Ursprung des Kunstwerkes,"* was a better understanding of what the artwork actually does: it unfolds the possibility of meaning and reference, of gods, dangers, etc., and also the possibility of taking a stand on the issues of its and our time. Finally, *Dichtung* came to be seen as the primordial art, because it sustained the closest and most fundamental relationship of any art to language. Through language, expressing the deepest beliefs of a people, *Dichtung* makes possible the opening-up of a historical world and thus the explicit self-recognition of that very people. *Dichtung* can be considered as "literature," if that term is widened to include "scriptures" and oral literature as well. Taken in that extended and yet fundamental way, it is clear that *Dichtung* participates vitally in the formation and preservation of "an historical humanity." Because Heidegger does not distinguish peoples on the basis of "race," or languages on the basis of phonological, syntactic, or semantic differences, *Dichtung* is able to emerge as a primary way in which an historical group of people recognizes itself as belonging together, as sharing a tradition and a destiny (*Geschick*).

In *"Hölderlin und das Wesen der Dichtung"* Heidegger follows this hint up with a further investigation of the nature of poetry (*Dichtung*) in the poet of the poet. Hölderlin's poetry (and his writing in general) possesses this twofold quality: it seems harmless, even eccentric (perhaps mad), and at the same time it draws out the very essence of language. It places the poet out in the "Between" of gods and men. In non-Heideggerian terms, the poet mediates the immediacy of language to the rest of humanity, he gives himself over to language and thus is able to free it from its appearance as means of communication or as game. The poet, bringing words to life, brings mankind before its own issues, those questions and decisions which decide who we are (*"wer der Mensch sei und wo er sein Dasein ansiedelt"* [HWD,47]).

As Heidegger reads Hölderlin poetizing the essence of poetry, the poet is not taken to be a god—i.e., he is not immortal and he does not possess extraordinary powers (as the Romantics would claim). The poet is part of humanity and a member of his people; like all human beings, he is born into a tradition. Unlike the mass of people, though, the poet manages (through his role and individual talent) to bring a wider group of people within range of the gods—i.e., of the "original" call of language. At the origin of a people, according to Heidegger, stand the gods as powerful and commanding figures. Their actions call forth a world and are considered important enough—indeed, "sacred"—to be spoken of, remembered, and re-enacted through poetry. Thus, Heidegger claims the gods are equiprimordial with the appearance of a world in the poetic word.

> . . . die Gegenwart der Götter und das Erscheinen der Welt sind nicht erst eine Folge des Geschehnisses der Sprache, sondern sie sind damit gleichzeitig. [HWD,40]

Whenever Heidegger speaks of time, we must read very carefully. "*Gleichzeitig*" is by no means "simultaneously" in the sense of "occurring at the same time." Rather, the gods draw near and the world opens up with the occurrence of language. They all thus share a time, they are coeval. Language is never just an empty formalism, it always sets up a world of concerns and relations. It must be remembered as well that "world" is not something man can have or lack; it is an essential part of his makeup, just as "understanding" (*Verstehen*) and "discourse" (*Rede*). Language is thus indispensable for all three.[6] And where language is active in its most fundamental sense is the poet's song.

After setting out the parameters of the artwork, and after bringing Hölderlin forward in the role of the "poet of the poet," Heidegger turns to consider *"Wie wenn am Feiertage..."*. The poem does not, however, simply present itself as so many words on a page, as it was taken in our initial reading (Chapter Two). Therein lies the particular value of Heidegger's reading. He does not allow us to simply "find" the poem, as we presumed in our immanent reading; it cannot be "unearthed," like an archeological artifact; it belongs to its "earth" as a river to its landscape. *"Indem das Werk eine Welt aufstellt, stellt es die Erde dar."* [UK,32] Nor is the poem accessible through purely philological means: no amount of elucidation of its textual basis, historical milieu, or biographical significance will attain it. It must first be sought as the kind of thing it is. Like the Greek temple Heidegger used in *"Der Ursprung des Kunstwerkes"* to exemplify how an artwork sets up a world, *"Wie wenn am Feiertage..."* itself sets up its own world and its own time, which makes possible the advent of the god (or in this case, of "the Holy").

This has not just been a temporal study of a poetic text. It has equally been a literary study of temporality. Literary methods were employed, not just to understand Hölderlin's poem, but also to approach Heidegger's writings as well. All too often, readers of Heidegger have overlooked his own style and method when trying to understand his words. At the same time, Heidegger has been quite damaging to traditional literary criticism. In each of the essays considered here, he has done an effective job of demolishing the pretensions of literary science to understanding or doing justice to poetic texts such as those of Hölderlin. It is difficult, after being exposed to Heidegger's thought, to return to reading poetry in the traditional way. It may even appear that the initial reading of *"Wie wenn am Feiertage...,"* in which the text was read for signs of temporality, was a waste of time. Surely Heidegger's reading superseded and replaced it. Undoubtedly, his way of reading the poem is incommensurable with standard philological practice. On the other hand, one could reject Heidegger's whole approach as hopelessly obscure and wrong-headed, and thus retain intact some version of literary-critical apparatus. In that case, the poem becomes "understandable" again, perhaps even laudable as a first attempt to write in a Pindaric style by a modern German poet. But does it not thereby lose precisely what makes it so valuable?

"Wie wenn am Feiertage..." is not read today merely as Hölderlin's first faltering step towards a growing mastery over the hymnic style. It is not merely of biographical or historical interest. On the contrary, it possesses and lives from its own power as poetic saying (*Dichtung*). No one, at any time, will replace or contain it. To list factors like its fragmentation, its stylistic experimentation, and the range of its "temporal signifiers" does not begin to touch the true importance of this text. What it says must be heard directly, without the intervention of analytic tools. It challenges the reader to hear in a way perhaps unique in literature.

At the same time, however, something remains to be said for continuing with a specifically "literary" analysis, even when approaching texts like *"Wie wenn am Feiertage..."* or *"Hölderlin und das Wesen der Dichtung,"* for it is as literature that they now speak, as carefully constructed works, each thoroughly embedded in its own temporality. Even Heidegger does not suddenly receive an insight and convey it directly (without mediation) to his audience. He too takes part in the temporality of his texts. He too must choose a path which leads through the texts in such a way that "thinking" can occur. He too does not remain unaffected by his readings. The hermeneutic circle, which we have followed so often in these pages, requires that reading be understood as a *process*. What distinguishes the poetic text from all others when read in this way is its *mode* of speaking.

Although Heidegger is plainly aware of the importance of the hermeneutic circle and of the process of reading as a tool for opening up the text to thought, many interpreters simply ignore his process in favor of arguing for or against his positions. They read Heidegger as abjuring the bare principle of method.[7] J. L. Mehta, for example, stresses the discontinuity in Heidegger's thinking:

> Heidegger realized that the whole conception of method is itself a manifestation and an integral part of ... subjectivism, that the leap by which alone thought can spring away from the sphere of essents to the dimension of Being must also leave behind the metaphysical concept of method . . .[8]

While it is true that Heidegger rejects method as a metaphysical *passe partout*, this does not entail that his own thinking lacks a method or that drawing it out would be useless. We cannot treat his words as if they were all surface, as if their meaning lay open for all to hear, provided only that one develops *"an ear that is sensitively and patiently attuned to the unthought and unuttered possibilities latent in the tradition of European philosophical tradition."*[9] I would suggest, on the contrary, that while Heidegger's process of reading certainly falls short of being a "method," in the scientific-analytic sense, it does play a significant role *on its own* in following the *"way to Being."*[10] Heidegger's frequent references to "path," "way," to following "hints" or "gestures" (*Winke*) show him to be acutely aware of the formal aspect of his practice of reading. As early as in *Being and Time* (1927) he demonstrated his great concern for the proper approach to ontological questions —as we saw in Chapter One—by a lengthy and carefully reasoned Introduction. Although he later dropped the designation of "phenomenology" or "hermeneutics," his way of investigating topics never lost its methodological character.

In the case of the present poem, we recall that Heidegger, after quoting three verses from St.II, asserts that

> Die innere Bewegung dieser drei Verse strebt auf das Wort "die Natur" zu und schwingt darin aus. Was Hölderlin hier noch "Natur" nennt, durchstimmt das ganze Gedicht bis in sein letztes Wort. [HH,52]

It is tempting to see such a passage—and it is indeed a critical one—as expressing Heidegger's main "thesis" or "argument" in *"HÖLDERLINS HYMNE."* He seems to be saying that a single word, *"Natur,"* is all we need to consider to understand the entire poem.[11] It would be the *Leitwort* of the whole text (both Hölderlin's and Heidegger's). To see it in this way, however, would mean ignoring both the process of the essay as well as the immediate context in which *"Natur"* appears. Heidegger here speaks of the "inner movement" of these verses; he thus distinguishes between what is evident on the surface and what may be active beneath that surface, a movement which highlights the word "Nature." His imagery, moreover, is aural: the inner movement "resonates" in the word Nature and "sets the tone" for the whole poem. It is thus invisible to analysis, it operates on the poem as a whole without necessarily showing up on its surface, which is to say: in its specific vocabulary or structure. Even this passage, which is obviously crucial for a content-based interpretation of *"HÖLDERLINS HYMNE,"* shows the insidious effects of Heidegger's style and his awareness of his own process.

Our own reading of Heidegger, however, has brought out the way he approaches *"Wie wenn am Feiertage...,"* reads through it linearly, only to return to the "center" of the poem (St.III, 1–2), not once but twice. Each reading of this central portion of the text has sought to deepen understanding of what has been read. In order to understand Hölderlin's importance for Heidegger, however, we had to read the essay in terms of its structure. What emerged was the crucial framing question concerning how or if Hölderlin's hymn had been heard. Heidegger heard it against the background of a world-historical decision facing the modern age. Why read Hölderlin today? he seems to say. For the simple reason that we need a poet like Hölderlin to even begin to understand what "today" means. Hölderlin is not writing merely for an audience in "1800," just as Heidegger is not speaking merely to an audience in "1939." If we restrict their saying to their own occasions, we will never hear them. Rather, it is clear that their time shares something with our time. Our time is not adequately "addressed" by reference to a calendar. Complicit with the time in which we live today are the background decisions which cut us off from yesterday. Only through consideration of those decisions could Hölderlin's poem be truly heard, and only through consideration of the structure of Heidegger's essay did the importance of the decision emerge.

Following Heidegger's process, however, only takes us so far; it cannot preclude a discussion of the substantive issues. It is certainly undeniable that, as most interpreters have insisted, *physis* or *Natur* is a central concept in *"HÖLDERLINS HYMNE."* Heidegger spends considerable effort to make clear his understanding of Nature as *physis*, a portion of which may repay careful reading here.

> Φύσις, *φύειν* bedeutet das Wachstum. Aber wie verstehen die Griechen das Wachstum? Nicht als die mengenmäßige Zunahme, auch nicht als "Entwicklung", auch nicht als das Nacheinander eines "Werdens". Φύσις ist das Hervorgehen und Aufgehen, das Sichöffnen, das aufgehend zugleich zurückgeht in den Hervorgang und so in dem sich verschließt, was je einem Anwesenden die Anwesung gibt. [HH,56]

Heidegger thus begins by interpreting a word occurring in the poem, *"Natur,"* as an unconscious resonance (*Nachklang*) of the Greek term *physis.* This word, he says, has nothing to do with "growth" as it is commonly understood. It is not a mechanical accumulation of empty ciphers, nor an organic process of development (*Werden*). *Physis* is rather the clearing of an open space, a *Lichtung*, in which the clearing itself recedes in favor of the entities it allows to come to presence.

Nature, in these terms, is the "forest" that is hidden by the "trees"—with the qualification that *physis* is no mere generalization from particular trees, but rather that which first makes possible the appearance of any trees whatsoever.

We may ask if this is an accurate description of *Natur* as it appears in the poem. Heidegger would reply, of course, that he is not seeking "accuracy" (*Richtigkeit*), but "truth" in the artwork. Even Hölderlin, according to Heidegger, hardly knew everything that he was saying with the word *Natur*.

> Aber Hölderlin hat die auch heute noch kaum ermessene Tragkraft des anfänglichen Grundwortes φύσις nicht gekannt. . . . Hölderlin dichtet in dem Wort "die Natur" ein Anderes, das wohl in einem verborgenen Bezug zu Jenem steht, was einstmals φύσις genannt worden. [HH,57]

Having said this, Heidegger returns to a close reading of the poem, though not in the sense of a literary *explication de texte*. Instead, he focuses on each word in turn, giving it the kind of interpretation suited to his purpose. He creates a kind of ancillary narration to elucidate that side of the poem which relates to Being.

There can be little doubt that Heidegger's reading of *"Wie wenn am Feiertage..."* disregards philological standards and seems willful at times. If "truth" is taken in the sense of "adhering to canons of correctness," then he clearly fails the test. But if it instead refers to the kind of truth discussed for instance in *"Der Ursprung des Kunstwerkes"*—truth as *"die Unverborgenheit des Seienden"* [UK,36]—then the attempt to measure the truthfulness of his reading must step back once again from a too-hasty judgement.

In ordinary parlance, truth is a correspondence between knowledge and fact. It is the correctness of a proposition. But, in order for there to be any "fact," any "thing" with which to correspond, there must already be an open space, *eine Lichtung*, in which it is revealed or accessible to us as anything at all.

> Das Seiende kann als Seiendes nur sein, wenn es in das Gelichtete dieser Lichtung herein- und hinaussteht. Nur diese Lichtung schenkt und verbürgt

> uns Menschen einen Durchgang zum Seienden, das wir selbst nicht sind, und den Zugang zu dem Seienden, das wir selbst sind. [UK,39]

With this view of truth, Heidegger asks to be judged by a far more difficult standard than empirical correctness. The truth of the work of art becomes evident by standing within it; how is that to be judged? In *"Wie wenn am Feiertage...,"* Heidegger is saying, Nature (*physis*) is the open space in which we have access to the kind of beings we are. Nature, *"wunderbar/ Allgegenwärtig," "mächtig," "die Allerschaffende,"* opens the clearing in which the poet feels called upon to speak, to "name" this event and thereby to "decide" the conflict between the *Wesen* and *Unwesen* of things [HH,58]. The poetic word, as living speech, thus gives form to Being.

Heidegger spends little time on the *Feiertag* in his reading. His reading of the opening strophe, in which the *Landmann* returns to his field, is similar to many others in its flatly descriptive tone.[12] Far more important for him is the appearance of Nature, which, as we have seen, he claims as the focus of not only the three verses referred to above (St.II,2–4), but of the whole poem from its beginning. The *Feiertag*, however, plays a central role in understanding the meaning of the reawakening of Nature. It is clear that neither the poet nor the gods nor the people can simply and arbitrarily decide the "time" for their coming-together. The time must be "right," the *kairós* must arrive. This is the time when Nature as the Holy reapproaches. Some critics (Paul de Man,[13] for example) have identified Heidegger's interpretation of this event with the *parousia*,[14] the emergence into presence of absolute Being. Such an interpretation, however, misses the significance of the temporality of such an arrival. Nature does not reawaken as though she were some being—even a divine being—who then could approach the poet as observer-subject. Such a description tries to fit the temporality of the poem into a framework of

> . . . the proximate end of the night of error by a cyclical return to the original illumination of truth, the temporal ecstasis that reopens upon a historical

> future conceived of in the form of an eschatology, and a closer adequation of destiny to Being.[15]

Temporality is not discovered through such a metaphysical system. On the contrary, metaphysics only serves to obscure the meaning of temporality in human life and reduces humans to mere playthings of time, robbing them of any other than the most passive role in its unfolding. To understand Nature's reawakening, we must think it in conjunction with the *Feiertag*.

The crux for understanding *"Wie wenn am Feiertage...,"* as we argued in Chapter Two, is the *Feiertag* itself. "*Feiertag*" names the time in which the *Landmann* returns to his land, and by extension it names the time in which the Holy approaches the poet. It is also the time in which the poet is cast into outer darkness by the gods. Is the poet thereby cut off from Nature? Impossible: Nature is "everpresent," she is always "coming back" from her absence. How then to understand the poet's fate? I believe this is only possible within a consideration of the temporality of the poem. Today is the day of days, the time out of time—the *Feiertag*.

Several studies of Hölderlin's poetry have shown how much he drew his conception of the poet and the function of poetry from his reading descriptions of Greek festivals[16] and his interest in the Pindaric hymns. Albrecht Seifert, for example, in his monumental study of Hölderlin's involvement with Pindar, has shown how the poet devoted much effort to translating many of the Epinikian Odes, in hopes of being able to transfer Pindar's style into German verse. Chief among Hölderlin's concerns was the temporal significance and function of the *Feiertag*. Seifert summarizes:

> Die Feier ist Lohn, Ertrag und Krönung der Arbeit. Auch von Pindar wußte Hölderlin, daß der Dichter die Feier mitzugestalten habe . . . Im Verhältnis der agonalen Bemühung zur Feier liegt nicht nur eine zeitliche Sukzession, sondern eine besondere Art der Simultaneität vor: im Fest als der letzten Station des Weges des Athleten ist der Kampf als Erinnerung aufbehalten.[17]

This very pregnant observation goes to the heart of the *Feiertag* in *"Wie wenn am Feiertage..."*. The *Feier* belongs to the victor as a living

Denkmal, which, because it exists in time, is temporary, but which also permits the recuperation of the memory of the struggle. The *Feiertag* in the poem, as this peculiar time, enables the poet to attend the return of Nature in the holy moment. And here we may suspect the presence of Hölderlin as existing human being behind the poetic voice. Ulrich Häussermann, while commenting on Hölderlin's later hymn *"Friedensfeier,"* also points to the central role of the *Feier* in Hölderlin's conception of the poet: *"Hölderlin hat seinen ‚Beruf' immer wieder mit dem Wort 'feiern' bezeichnet."*[18] If Hölderlin is as Heidegger calls him, *"der Dichter des Dichters,"* then he is so as a *celebrant*.

To understand the significance of the *Feiertag* in this poem, it helps to review the historical background of the word, much of which would surely have been familiar to Hölderlin.[19] The *Feiertag* is a kind of time. Human beings live in time, recognize time as a factor in their lives, seek to benefit from, save, or maximize time, etc. Time as such, however, is not homogeneous. It might seem that each minute, hour, or day is like every other, but even in this abstracted framework (which is only found in modern technological societies), it is clear that some days are more important than others, that some minutes seem to take hours, etc. In all cultures, moreover, certain times are set aside for special activities. As Hesiod knew, certain activities are "right" to engage in at certain times.[20] Work, play, love, rest—all only take shape in their times. The *Feiertag* is such a time. The times for the earliest celebrations, at the turning of the year, returning of seasons or crops or animals, were often based on naturally-occurring phenomena. By observing these phenomena, earlier peoples gave expression to the belief that only actions taken in "proper" times could have any assurance of success.[21]

The term *kairós*, which we have used frequently in this study, is usually translated as *"exact or critical time, season, opportunity."*[22] Recent scholarship, however, suggests that up to Pindar's time, it was not primarily temporal; it was closer to in meaning to "due measure."[23] The *kairós* would belong to the whole constellation of warnings about "excess" or transgressing boundaries in Greek literature. Time was thus not uniform, as we think of it, but closer to the model of fruit ripening "in the fullness of time." It is interesting to note that even peoples like the ancient Greeks—who lacked calendars as we know them and such concepts as "week," or a fixed chronology in "years"[24] (measured by revolutions of the sun), and whose "months" were irregular and inconsistent among the various city-states—distinguished carefully between favorable and inopportune times. Such discrimination could obviously be essential for survival in a harsh terrain. With increased astronomical sophistication, people were later able to rationalize and even predict the arrival of these favorable times. They always remained essentially independent of man, however. When the time for planting or harvesting, for making war or making peace, would arrive was not predictable. People at that time were in the position of the *Landmann*, who must "wait and see" (*"ich harrt ... und was ich sah..."*) what the gods brought and *when* they brought it.

The *Feiertage* themselves, then, were "gifts" of the gods, a re-presentation of the original benefaction which they were intended to commemorate. The *Feiertag* marked the time in which the god performed his work and thus it was dedicated, or "sacred," to the god. By clearly distinguishing itself from ordinary time, which was given over to humans, it helped maintain the separate but related identities of men and gods. Often—as through sacrifice—it underlined the role of human beings in keeping the cosmic order in balance.

Clearly, the *Feiertag*, or holiday, or "day of celebration," is a special time, and is observed in special locations which are marked off from the ordinary. *"Die Orte und Zeiten sparen aus den alltäglichen, zweckhaft dienlichen Räumen und Zeiten besondere Bezirke für das feierliche Geschehen aus."*[25] Its periods of arrival, duration, and departure were (and are) carefully delimited and dependent on the observation of natural phenomena—the appearance of birds, animals, plant cycles, etc.[26] Such careful delimitation was necessary to avoid the possibility that the extraordinary time, in which ordinary rules of behavior were set aside, did not extend itself without limit and thereby bring destruction in its wake.[27] The celebration of the *Feier* thus also served to preserve the links of the community with the rhythms of Nature and the gods.

Most importantly for our discussion of *"Wie wenn am Feiertage...,"* the *Feiertag* called attention to the nature of time itself, through its connection with naturally-occurring phenomena which were present at some times and not others. Whereas in secular time every day might seem like every other, the *Feiertag* marked a new beginning on the cycle. It made clear that all times were not the same, and it underscored the importance of correctly remembering and commemorating specific times, particularly those marking the beginning and end of a cycle. As Wolfgang Binder has written, in reference to Hölderlin's concern with the temporality of the *Feier*:

> Die Feier ist also in einem doppelten Sinne ein zeitlicher Vorgang. Sie läßt die verdichtete Gegenwart des Jetzt, gleichsam die Akmé der Zeit, entstehen, und sie wiederholt im feiernden Gedächtnis ein früheres Ereignis. Damit stellt sie über eine leere Strecke hinweg die Kontinuität von Gipfelstunden und so in der verfliessenden [sic] Zeit die Ständigkeit und Fülle eines höheren Lebens her.[28]

The double function of the *Feiertag*, as Binder describes it, sets the present, as a special and consecrated NOW, into a continuity with the remembered past. Indeed, it is through the celebration that the past is kept alive in memory, although usually in altered form. The celebration brings together and consolidates the community in a special moment which is not, however, celebrated for its own sake (to have a good time), but to maintain links with the tradition of the community.[29] Speaking of the Dionysia, but applicable to other festivals as well, Christian Meier writes,

> In diesem Fest erfuhr sich die Bürgerschaft als Einheit . . . Sie tat dies in der Erfahrung ihrer Menge, ihrer Gemeinsamkeit, aber auch im gemeinsamen Erleben der Stücke, der Erregung durch das Geschehen, der Spannung, die den Aufführungen vorangegangen war, die sie dann beherrschte.[30]

No celebration would be complete without singing and dancing, and quite early an individual was designated to lead the songs. That person was the poet. Nor was he merely the conductor or prompter of the chorus, someone who essentially stood outside its singing. Rather, his role was specifically to cast the ancient hymns into new words. Walter Burkert describes the celebration of Greek festivals in classical times thus:

> Although the names and basic rhythms of the dances are traditional, the cult in no way demands the repetition of ancient, magically fixed hymns. On the contrary, the hymn must always delight the god afresh at the festival; there-

> fore for dance and hymn there must always be someone who makes it, the poet, poietes.[31]

The *Feiertag* (from Latin *feriae*[32]) is the day on which the poet truly steps into his own role and leads the people in celebration. Indeed, the poet is the only individual distinguished from the mass of celebrants (both by Hölderlin and the Greeks) and thus takes on a representative function in approaching the god. In this he acts similarly to (and often in conjunction with) the priest. *"Wie wenn am Feiertage..."* certainly illustrates the religious-representative aspect, as, for example, when the poet proclaims:

> Doch uns gebührt es, unter Gottes Gewittern,
> Ihr Dichter! mit entblößtem Haupte zu stehen,
> Des Vaters Stral, ihn selbst, mit eigner Hand
> Zu fassen und dem Volk ins Lied
> Gehüllt die himmlische Gaabe zu reichen. [St.VII, 3–7]

The *Feiertag* in *"Wie wenn am Feiertage..."* was only mentioned once, at the opening of the poem. It was not even "present," in the poetic NOW, since it was part of the opening hypothetical description (*"Wie wenn . . ."*). Nevertheless, more than any other single factor, the *Feiertag* enabled the poem to say what it had to say. It underscored the special nature of the time in which these events were taking place. *"As on a holiday..."*—thus, *not* on an ordinary day, a *work*day. Because it was a *Feiertag*, it set the "time," so that the *Landmann*'s going to see the field took on the significance of one returning to the ground of his Being and *seeing* the whole, the relationship of micro- and macrocosmos, for the first time. On an ordinary day, his own concerns, worries, and plans would have rendered the field invisible. By going out on the *Feiertag*, the *Landmann* placed himself in the necessary *temporal* frame to be able to see what his world looked like, as it was in truth. To do this, he needed the *Feiertag*: the "time out," sanctioned by the gods.

Readers have often confused the *Landmann* with the poets, who are the true subject of the comparison. A careful reading of these lines shows, however, that the poets are actually in the position of the *"Bäume des Haines,"* who have weathered the nocturnal storm and now stand *"unter günstiger Witterung."* They thus set up the comparison with Bacchus in St.VI, by having weathered the storm (Zeus), and—together with the vines—they offer humans a safe place (*Hain*) for encountering the divine. The *Landmann*, I would suggest, is actually in the position of the reader of the poem.[33] Without him, the whole comparison would lose its point. (Does a tree falling in the forest without someone to hear it make a sound?) Nature does what it always does; the trees, vines, and stream do what they always do—only the *Landmann* steps out of the ordinary, "natural" order of things—and *sees*.

The dilemma which faced our reading of the poem in Chapter Two was the obvious incompatibility of the kind of time described through the first triad, and indeed throughout most of the poem, with the kind described in the last triad. The poet seemed to struggle to maintain the favorable aspect of the time for poets, yet could not refrain from speculating what would happen "if..." (*"Denn* sind nur *reinen Herzens* ...sind *schuldlos unsere Hände...Doch weh mir!* wenn von...*Und* sag ich *gleich..."*). The conclusion seemed inescapable that the poetic voice itself was split, fragmented between the beginning and the end of the poem.

It seems the urge to explain or harmonize this cleft in the poem is irresistible. Virtually every interpreter has attempted to find some way of reconciling the apparent self-castigation of the final strophes with the rest of the poem—Heidegger being one of the few exceptions.[34] The challenge has been, it seems to me, to read the poem without either ignoring its fragmentation or harmonizing its contradictions and thus depriving it of a source of its power. To read the poem adequately, I would maintain, requires an understanding of the *Feiertag* such as has been presented above.

The *Feiertag* is important because it 1) removes man from "ordinary" time, 2) brings him back to the basis of his existence, 3) is inextricably bound to song (*Dichtung*) and thus to the poet, and 4) calls for recognition of the DECISIONS which have framed the poet's (and his community's) world. The *Feiertag* is not within man's power to control; it comes in its "own" time. Not even gods can influence the rhythms of the *Feiertag*. It thus demands recognition as a superior power over humans. Yet its arrival does not obsolve men of responsibility but rather calls forth recognition of humanity's own involvement in its past and future.

In the poem, the *Feiertag*, like the *Landmann*, appears only once and then vanishes from the text. It thereby makes way for the actual events that are to take place: the retreat of the storm, the reawakening of Nature, the enthusiasm of the poets. Its withdrawal enables these things to come forward and be seen.[35] It withdraws before the actual celebration and its multifarious appearances, although it is the unspoken subject of the whole poem, since the *Feiertag* is what the poets celebrate, and since Semele's story is obviously meant to illustrate the origin of the *Feiertag* (and of the poetic role). Thus, after beginning with an ontological perspective, the poem returns to the everyday, the very surface of life. The *Feiertag*'s function of transforming time allows the specific events of celebration to come forward, to take place and to be understood as what they are. It permits certain behaviors which would otherwise be prohibited (sacrifice, revelry, etc.) and prohibits other behaviors which are otherwise considered quite "normal" (e.g., work or war).[36]

In this way the poet in *"Wie wenn am Feiertage..."* proceeds in a similar fashion to the hermeneutic circle as we have followed it in Heidegger's readings, only in the opposite direction. Whereas Heidegger's investigations began with things in their everyday and unreflected appearance, only to step back in favor of their fundamental Being, the poet has begun with the *Feiertag*, which proclaims the basis of human temporality, only to withdraw before the ephemeral manifestations of Nature. It must not be inferred, however, that the "ephemeral" is somehow less valuable or less real than some more prolonged phenomena. All of the celebrations that take place on the *Feiertag* only take on meaning through their consecration to that one day (regardless of how long the "day" lasts).

It is through the *Feiertag* that the poem is able to address the issues that Heidegger finds so important. The place of the poet, the role of Nature, the danger of forgetting Being, all derive from the presence or absence of the *Feiertag*. The *Feiertag* is the day when the poet comes into his own, when he commemorates through the ritual enactment of the festival the acts of the god(s) and thus founds anew the time of the community. By affirming their relationship to the gods, the people identify themselves, they develop and maintain a sense of their origins and destiny. The *Feiertag* sets the parameters under which the people recognizes itself in its world. Burkert writes, *"As the sanctuary articulates space, so the festival* [i.e., *Feiertag*] *articulates time."*[37] To be allowed to take part in the *Feier* signifies acceptance into the community and recognition of that community's self-definition, its history.[38]

Heidegger, in the later pages of his essay, speaks extensively about *"das Heilige."* What he does not acknowledge, however, is that the poet's function of mediating between the Holy and the people is only possible through the arrival of the *Feiertag*. Only then can the poet truly say, *"Das Heilige sei mein Wort."*[39] To speak of it at some other time would be blasphemy, or at least wasted breath. During normal times, the word of the poet is not heard, it is drowned out by the sound of work. Only in the stillness and *Feierlichkeit* of the *Feiertag* does the poet's word assume its true significance. Even in scenes of utmost revelry and abandon, the moment arrives in which stillness descends and the day is consecrated. It is in that moment when the poet can be heard. Rudolf Nießen, writing about the experience of the *Feiertag,* explains that it is

> ...immer der Einbruch eines Unaussprechlichen in das konkrete menschliche Dasein. Dies vollzieht sich im Augenblick, in dem die Zeit stille steht und der aus der Stille geboren ist.[40]

The casualness with which Heidegger skims over the *Feiertag* in *"HÖLDERLINS HYMNE,"* on his way to an exposition of *Natur* as *physis,* suggests that he overlooked it. He did not forget it, though, since in his lecture series on Hölderlin's *"Andenken,"* held in 1941–2 (i.e., at roughly the same time as *"HÖLDERLINS HYMNE"*), he returns to the *Feier* in several sections.[41] He begins with a caution (perhaps an excuse) that—even in remarks on *"Wie wenn am Feiertage..."*—it might be necessary precisely to keep quiet about the "decisive word" (*das entscheidende Wort*) in the first verse.[42] What he says about the *Feiertag* in the lecture about *"Andenken,"* however, has a bearing on our discussion here. In that context, Heidegger does indeed recognize the *Feiertag* as decisive for the history of a people. Although in ordinary language it is usually described negatively, as the cessation of work, a pause in ordinary time, Heidegger sees that the purpose of the *Feiertag* is actually to free humans *from* their captivity in the ordinary and to free them *for* authentic time.

> Feiern ist das Freiwerden vom nur Gewöhnlichen durch das Freiwerden für das Ungewöhnliche. Feiern ist Aufhorchen auf das, was Stifter das "sanfte Gesetz" nennt, ist Erwartung des Eigentlichen, ist Vorbereitung der Aneignung des Wesenhaften, ist Erharren des Ereignisses, in dem das Wesenhafte sich offenbart. [A,66]

The *Feiertag* is the day of the festival; the festival can only appear in the time set aside for it. Both depend on each other. This is a time of renewal as the community comes together and celebrates the occurrence of the approach of gods and men. *"Das Fest ist das Ereignis des Entgegenkommens der Götter und Menschen."* [A,69] How do they perform this event? By staging a performance: a "tragedy" (*Trauerspiel*):

> Die "Trauerspiele" der Griechen sind Festfeiern. Das sagt: Hier fallen Entscheidungen über die Entgegnung und aus der Entgegnung der Menschen und Götter. Somit wird jedesmal auch über den Menschen und über die Götter entschieden. Die Feste sind die eigentlichen Geschichtstage, will heißen: geschichtebildende Tage. [A,78]

So once again we return to the crucial element of the temporality of the *Feiertag*: it makes possible the decisions which define the relationship of the people to their gods. At the same time, it is constituted by the encounters of the people and their gods. It is the scene of transformation of the people's history because it defines what truly matters. The only word possible for this sacred function is: *das Heilige*.

> Das Festliche des Festes, ... ist das anfängliche Grüßen *des* Grüßenden, das Hölderlin in der ersten seiner Hymnen "Wie wenn am Feiertage..." "das Heilige" nennt. [A,70]

With this gesture, Heidegger recuperates the discussion of "the Holy" which he held in *"HÖLDERLINS HYMNE."* Although Heidegger there does not speak directly about the *Feiertag*, he does recognize that the relationship of the poets to the Holy carries danger with it. Contrary to almost all other interpreters of *"Wie wenn am Feiertage,"* though, Heidegger sees the danger directed against the Holy itself, rather than against the poets. They are, after all, sanctioned by the gods to speak of the Holy on these "holy days." But what is to guarantee that their words will do justice to the occasion? What, indeed, is to prevent people from taking the poet's words themselves as "the Holy?" Then truly the Holy, which is the immediate, would be transformed into its opposite.

> Allein, gerade dieses, daß das Heilige einer Vermittelung durch den Gott und die Dichter zugewiesen und in den Gesang geborgen wird, droht das Wesen des Heiligen in sein Gegenteil zu verkehren. Das Unmittelbare wird so zu einem Mittelbaren. [HH,72]

For that reason, the poets are specially chosen from the mass of the people to approach and to remain with the Holy. By remaining with the god, they partake of his *Leiden* (*pathoi*): *"die Leiden eines Stärkeren /Mitleidend,"* in which *Leiden* is not "suffering" so much as the concentrated experience of what it means to be "in" the world (*Innigkeit*). The poets are thus called to remain at the beginning (*"Das Leiden ist das Festbleiben am Anfang"* [HH,75]). They, above all, must not relinquish the source of the community's identity.

At the end of his *Erläuterung*, Heidegger returns to the third strophe and asks again about the NOW: *"'Jezt'—wann ist dieses 'Jezt'?"* He answers that it is the time of the coming of the Holy, when "history opens itself up to essential decisions" *("die Geschichte sich wesentlichen Entscheidungen stellt").* By connecting the Holy with the *Feiertag*, Heidegger reëstablishes its temporality. The Holy is not an immutable, everpresent *eidos*; it is active, approaching and receding from the people. It is therefore imperative to grasp it when it comes. Such times consecrate the present and make history and historical decisions possible.

We have seen how the poem itself presents the *Feiertag* as the time of the approach of the Holy. Viewed naturalistically, it would appear to be spring, when storms and sunlight bring growth to the fields. Hölderlin, however, seems to be talking about a larger season, a season of the "year" in which human history is viewed cyclically. At such a turning point, the *Feiertag* comes around and the poet steps forward to sing of "the Holy": i.e., the indissoluble (and therefore "whole") relationship of gods and men, of generations.

The *Feiertag* celebrates the coming of the Holy as a time in which historical decisions are recognized as "still outstanding"; i.e., as needing to be made. We must now take stock of what has been learned about the nature of the "decision" in the course of reading Heidegger's essays.

Heidegger has called Hölderlin the *"poet of the poet"* because he *"places us before the decision."* [HWD,34] We, the hearers and readers of the poem, are those who must reach a decision. The decison involves *"who man is and where he settles his existence."* [HWD,47] Hölderlin's word *"says the Holy and names thus the unique time-space of the initial decision for the essential structure of the future history of the gods and humanities."* [HH,77] Stringing Heidegger's words together in this way carries with it the danger of making them all seem to be part of the same discourse. Perhaps in a sense they were; however, they all appeared in separate essays, in separate circumstances. What they are meant to demonstrate here is the interconnection between Heidegger's interest in Hölderlin's verse and the importance of the poet for articulating the "decisions" facing humanity. Specifically, in what kind of time will man dwell: will it be with or without the gods?

To approach Heidegger's understanding of the decision called for by Hölderlin, we must first look at the extraordinary language of *"Wie wenn am Feiertage..."* itself. Some modern commentators have found an explanation for the peculiar style of Hölderlin's hymns in the historical circumstance that, unlike Pindar, he did not actually compose for a public festival: he could not in fact expect to step forward and lead his people in hymns of praise to the gods. Because Hölderlin was "out of step" with his time, according to this interpretation, he was spurred into ever further extremes, both in verse and in life.

> The world of Pindar was in all respects more favourable to poetry than that in which Hölderlin was fated to live. . . . [T]he festivals he celebrated existed only in his hopes.[43]

Unlike the situation in contemporaneous France, where "revolutionary festivals" were actually being held, where hymns to ideals like *liberté, égalité, fraternité, la paix, la patrie*, etc. were indeed sung in public,[44] Hölderlin composed his poems knowing very well that they would only be read, if at all, by individuals. In fact, they would be almost impossible to "sing" chorally as the Pindaric hymns had been.

Although it would seem obvious that the lack of audience would affect Hölderlin's mental state while composing his works, I cannot go so far as to ascribe their character to this fact. It remains undisputed that he did choose to write such hymns in full knowledge of the absence of an audience. While he may have hoped for the arrival of such an audience and of such festivals at some unspecified, but future, date, he clearly recognized by 1800 that any such events would have to be instituted by individual poets and their readers.

Heidegger, on the other hand, recognizing the idiosyncratic status of *"Wie wenn am Feiertage...,"* has taken the poem seriously as the kind of hymn it is: not a public choral *Gesang*, but a poetic text in the form of a hymn which itself constitutes the *Feier* it seeks to celebrate.[45] Heidegger, in *"Hölderlin und das Wesen der Dichtung,"* claimed for the poet the foundational role of *"naming things to what they are."* [HWD,41] The poet sets up the conversation between gods and men and thus brings time to language. All time is based on discrimination: past, present, and future must be set off from one another in some way. The poet, in creating or bringing forth words that cannot be forgotten, words that demand rehearing, whether because they relate the wondrous deeds of the gods or because of their own "wondrousness," lays the foundation of language, as "that which remains" (*das Bleibende*). Such words call forth time and history through remaining or enduring in time and history.

> Erst seitdem die "reißende Zeit" aufgerissen ist in Gegenwart, Vergangenheit und Zukunft, besteht die Möglichkeit, sich auf ein Bleibendes zu einigen. *Ein* Gespräch sind wir seit der Zeit, da es "die Zeit ist." Seitdem die Zeit aufgestanden und zum Stehen gebracht ist, seitdem *sind* wir geschichtlich. [HWD,39; italics in original]

It is clear to every reflective person that there are issues "in today's world" that are waiting resolution. How are these issues made accessible to us? Through various media, all of which ultimately derive from language. Even images can only engage in discourse through language. Where does language itself find words with which to frame the issues that are so important to us? Sometimes words can be traced to a specific individual, sometimes new coinages simply "appear" and gain currency. Speaking in terms of the eighteenth century, one can ask where the idea of the "rights of man" came from, one can track down references to it in texts. Its origin, however, is properly speaking "lost." At the same time, it is clear that it must have had an origin somewhere, someone must have first formulated the idea. Heidegger points to the creative agency of the poet in bringing words (in the form of issues) before us, so that we may have the means of discussing, analyzing, and—possibly—thinking about them. The issues that face us will always have been inaugurated by poets. They who have heard the original voice of language have carried it over in song (*"ins Lied gehüllt"*) to the rest of us.

The poets bring us before the "decisions" of our time. They do not necessarily pose questions or demand answers. The decisions are far more fundamental than that they should be encompassed by rational discourse alone. The decision, for example, whether we will live with the gods or in their absence is not up for debate (although often debated).[46] Before the question can even be asked, the answer has been given. When the gods are near, their existence is not questioned. When they have departed, their absence is palpable, although many may seek and even "find" them. The poet does not offer solutions, he is not a scientist, detective, or doctor. Rather, the poet speaks the truth, he does not let us forget. By thus *"putting the truth to work in the work,"* he brings us before the decision.

Although I have claimed that the concept of the "decision" was important for understanding Heidegger's view of Hölderlin and indeed for understanding the "essence" of poetry, Heidegger himself does not spend much time in the essays we have considered elaborating the concept of the decision. It might be of value to look briefly at a text in which he addresses the issue of the decision more explicitly.

Sections 43–49 of *Beiträge zur Philosophie (Vom Ereignis)*[47] deal with *"Entscheidung"* in a variety of modes. Many of the sections seem to be no more than notes jotted down for future use (§§46,48,49). Sections 43, 44, and 45, however, which are the longest of the group, go into some detail about how Heidegger understands the term.

In §43 Heidegger discusses the danger of reading *Being and Time* "anthropologically" and of seeing the connections between *Entschlossenheit/Wahrheit/Dasein* as being determined by *Entschließung*, instead of arising from *"dem waltenden Grund des Da-seins."* He emphatically places the term *Ent-scheidung* in the center (*"die innerste Wesensmitte"*) of *Seyn* itself.

He distinguishes clearly between a psychological understanding of "decision" (*Entschließung*) and "decision" (*Entscheidung*) as *he* understands it. The first kind of decision involves choice, "deciding" (*Entschluß*), and preference; the second kind is situated in its *seinsgeschichtlichem Wesen*. He points to the danger of interpreting *Entschlossenheit* against the background of moralistic *Entschließung* instead of *Entscheidung*. He defines the latter as

> ...das Auseinandertreten selbst, das scheidet und im Scheiden erst in das Spiel kommen läßt die Er-eignung eben dieses im Auseinander Offenen als der Lichtung für das Sichverbergende und noch Unentschiedene, die Zugehörigkeit des Menschen zum Seyn als des Gründers seiner Wahrheit und die Zugewiesenheit des Seyns in die Zeit des letzten Gottes. [BP,88]

Since I think this passage is not easy to understand even at the level of syntax, I would like to suggest the following translation as a basis for discussion:

> [The de-cision says] breaking-apart itself, which splits and by splitting first brings into play the "occurrence of appropriation." [It is the occurrence] of just this openness-in-apartness [which is] the clearing for that which hides itself and has not yet been decided, [i.e., that] humans belong to Being (*Seyn*) as to the founder of their truth, and that Being has been assigned to the time of the last god.

Heidegger's discussion of the decision here is circular: the decision brings forth the split which allows for something to be outstanding and "not yet decided." It is not a matter of "making a choice" (*das Treffen einer Wahl*). The de-cision is a splitting or breaking apart which sets the stage for man. By this splitting, the de-cision allows things to step forward as matters of concern, matters for "decision" (choice). He sees man belonging to his "truth" and to his "time," which is to say, to that which founds meaning and that which overthrows it. Man belongs to a particular disclosure (de-cision) of Being, which makes possible a particular "truth." But that truth carries within itself its own "time," i.e., its own finitude. He seems to be referring to the foundational split between the ontic and the ontological.

The de-cision opens the realm of possibility where a difference can appear, make sense, and actually *matter*. A political state or nation, for instance, may have traditionally seen itself as essentially homogeneous. If, at some particular "time," a rupture occurs and people begin to see themselves primarily in terms of adherence to separate (ethnic, economic, social, regional, educational) groups, we could speak of a fundamental "breaking-apart" or "de-cision" occurring. Only then would issues such as "independence," "self-determination," or "discrimination," become important. The initial breaking-apart would have let these issues step forward as awaiting decision: they would be *"das noch Unentschiedene."* I would like to emphasize that they need not be explicitly formulated as "issues," they do not necessarily have a program or rationale. Rather, it is always self-evident that "people are just like that," or simply that "something must be done." This example also brings forward how decisions not only occur in time but actually determine the character of their time by setting its parameters. If the decision is fundamental enough it may someday come to be fixed as a *Feiertag*.

We find in §45 that Heidegger again uses the term *"er-springen,"* which earlier had appeared in *"Der Ursprung des Kunstwerkes,"* where it described the nature of an "origin:" *"etwas erspringen, im stiftenden Sprung aus der Wesensherkunft ins Sein bringen, das meint das Wort Ursprung."*[48] He defines the term "*erspringen*" as bringing something out of its essential tradition into Being, through the "founding leap," i.e., as an *Ursprung*. Here it describes how we "get" a de-cision: not through choosing, willing, or deciding, but by letting something become an origin for us.

Neuzeitlich gesonnen denken wir von uns aus und stoßen, wenn wir von uns wegdenken, immer nur auf Gegenstände. Diesen gewohnten Weg des Vorstellens eilen wir hin und her und erklären in seinem Umkreis Alles und bedenken nie, ob nicht dieser Weg unterwegs einen Absprung zulasse, durch

> den wir erst in den "Raum" des Seyns springen, *uns die Ent-scheidung erspringen.* [BP,88; italics added]

Heidegger in this passage attacks the subjectivistic way of reasoning which starts with the Cartesian *ego* and proceeds, in truly "egoistic" fashion, to come upon "objects" outside, which it then attempts to explain exhaustively. Such a "modern" (*neuzeitlich*) approach will never approach Being thoughtfully. The de-cision does not get "decided" (*getroffen*) in the normal sense as much as it is made through a "leap into Being," the way one can either "*make* the grade" or not. This leap has the character of an origin. An *Ur-sprung* is not the completion of some act but the *Auftakt* which discloses new possibilities for man.

In order to clarify his understanding of the term, he suggests looking at some *Entscheidungen* which have emerged as "historical necessities" from the original *seinsgeschichtliche Auslegung der Entscheidung* (which he designates by the form *"Ent-scheidung"*). This makes up the content of §44, titled *"Die 'Entscheidungen,'"* which provides numerous examples of the kinds of decision Heidegger means. Some seem to place responsibility directly on man's shoulders; e.g.,

> [Die Entscheidung] ob der Mensch überhaupt noch die Entscheidung wagt *oder* ob er sich der Entscheidungslosigkeit überläßt, die das Zeitalter als den Zustand "höchster" "Aktivität" nahelegt. [BP,91]

> [The decision] whether man is still capable of facing the decision at all *or* whether he relegates himself to doing without any decision, which our age regards as the condition of "highest" "activity."

It is important to notice that the decision Heidegger here poses is not between two symmetrical alternatives, but between the freedom to make a decision or the lack of it. Every historical people lives within and out of the decision it reached sometime ago, "at the beginning of time," a decision which defines *who* that people is and which opens the possibility of having meaningful goals. But such a decision carries enormous burdens of responsibility with it, and now, in the age of technology, it would seem so much easier to simply do without any decision at all. Of course, that would entail doing without history as well. The place of history could be taken over by a "historical science," in which all origins would be "equal," but which would never discover a path connecting people now alive to their ancestors. Bereft of even the ability to recognize such decisions, people could occupy themselves with "technical" questions, involving means without consideration of ends. Such "activity" characterizes lostness in *das Man*, where one follows the general, public alternatives in an area of life, without taking them over authentically.[49] *Das Man*, or "the One," as Hubert Dreyfus translates it,[50] expresses Heidegger's view that Dasein exists for the most part in public, inauthentic ways, which, however, provide the basis of intelligibility of its world. Dasein, in its everyday aspect, understands what it finds in its world, including itself, simply as "one" does. It loses itself in frantic activity, which is actually an *Entscheidungslosigkeit*, or a state of being bereft of any decision.

Each of the decisions is offered as "whether...or..." (*"ob...oder..."*); i.e., as laying out the way Being has been carved up (*ent-schieden*) in each case. Several alternatives in §44 recapitulate issues from *"Der Ursprung des Kunstwerkes,"* including the decision

> . . . ob die Kunst eine Erlebnisveranstaltung *oder* das Ins-Werk-Setzen der Wahrheit ist.

And:

> . . . ob die Natur zum Ausbeutungsgebiet des Rechnens und Einrichtens und zur Gelegenheit des "Erlebens" erniedrigt wird *oder* ob sie als die sich verschließende Erde das Offene der bildlosen Welt trägt.

Each of the "decisions" clearly requires an exclusion as well as an affirmation. As something which "splits," the de-cision opens up a realm where beings can belong together through conflict (*Streit*). But Heidegger does not openly engage in polemics here. There is no talk of moral imperatives, although it is hard to believe each of the alternatives is equally attractive to him personally. They are simply "historical necessities:" they are part of the world into which Dasein is thrown, they have been given to him as something "outstanding" with which to be involved.

Although they appear quite varied, all of these decisions come down to one: whether *das Seyn* withdraws for good or whether its (current) withdrawal can become the "first truth" and the "other beginning" of history.

> Alle diese Entscheidungen, die dem Schein nach viele sind und verschiedene, ziehen sich auf die eine und einzige zusammen: ob das Seyn sich endgültig entzieht *oder* ob dieser Entzug als die Verweigerung zur ersten Wahrheit und zum anderen Anfang der Geschichte wird. [BP,91]

The withdrawal of Being is seen as opening the possibility of a new beginning of history. History had previously been seen in terms of "positive" truths, when the gods seemed to live near man, when the world was in order, or when at least utopia was realizable. It is difficult to conceptualize the *"irreversible withdrawal of Being"* unless as the total annihilation of mankind. The alternative would be to found a new history on the basis of the inaccessibility of Being.

In everyday language, "someone" makes a decision. He or she thereby becomes a "decision-maker," which is the current euphemism for someone with the power, authority, and responsibility for "making things happen." Heidegger does not think the decision that way. He asks,

> *Wodurch fällt die Entscheidung?* Durch das *Geschenk* oder den *Ausbleib* jener ausgezeichneten Gezeichneten, die wir "die Zukünftigen" nennen im Unterschied gegen die vielerlei beliebigen und unaufhaltsamen Späteren, die nichts mehr vor sich und nichts mehr hinter sich haben. [BP,90]

No ego possesses the power to "make" the decision here. Rather, certain individuals are either "given" to us or "withheld" from us: *and the decision has already occurred* through the giving or withholding of that gift. We might think of these "future ones" (*"die Zukünftigen"*)—whom even Heidegger places in quotation marks—as "the right person at the right time." We cannot say where they come from or what they look like; they simply appear. They represent something new, something not covered by old ways of being or thinking. Merely proceding with the status quo is no longer enough, and these individuals render that inadequacy undeniable. The decision is found when the present reaches into the future in a significant way, instead of being cut off from reference to anything outside itself. Heidegger also does not say that an individual makes such a decision; or even that a people makes it. He merely speaks of "preparing" the decision, which is not the same as preparing "for" it; he does not suggest a psychological readiness for a decision, but says that the decision itself must be prepared: i.e., it must be properly laid out. *"Wie kann diese Entscheidung vorbereitet werden? Haben hier Wissen und Willen einen Raum der Verfügung, oder würde das nur ein blinder Eingriff in verborgene Notwendigkeiten?"* [BP,97] These questions are not directly answered.[51] Instead, he proceeds to talk of the "need" (*Not*) within which decisions must be faced. *"Aber Notwendigkeiten leuchten auf nur in einer Not."* This *Not* threatens to transform the opportunity of preparing the decision into merely a hastening of the growth of ahistoricality. *"Wer um* diese *Not nicht weiß, ahnt nicht einen Schatten von den bevorstehenden Entscheidungen."*[97] It is clear that history may present us with the necessity for a decision, but that it cannot supply the decision itself. History supplies the preconditions for a decision which the decision dismantles for its own purposes. In that respect, history must be "forgotten" in order for the decision to be made.[52]

In the passage cited above, Heidegger seeks to distinguish individuals (*Einzelnen*), cohorts (*Bündischen*), and those who are simply stuck with each other (*Zueinanderverwiesenen*). This, he claims, is the constitution of a *Volk*, in which most people are together simply because they happen to have been born in a particular time and place. All members of a people are related to each other by their common history, but that is still at best a distant relationship. And yet, it is only through and for the sake of these anonymous "compatriots" that the truth of the whole people emerges. With some members of the community, however, there is an additional kinship: they are related by family ties or common values. With them we realize the aims of the few. Finally, there are those rare individuals who clear the path for the future development of the people as a whole. They found in advance (*vorausgründen*) the time and place for a new disposition of Being.

This passage could obviously lend itself to a nationalistic interpretation, although Heidegger does not give it one and in fact seems to have quite different concerns. The decision being discussed here is not the *result* of so-called "world-historical events," as is usually assumed.

Rather, the decision itself must create its own "time-space" in which the highest seriousness and greatest joyfulness come together in the "essential moments" (*wesentliche Augenblicke*). The danger arises, however, that the decision can be made which reduces even "culture" and "*Weltanschauung*" to mere weapons in the battle for naked survival of the "*technisierte[s] Tier.*" [98]

The de-cision, as Heidegger uses the term, speaks of the articulation of Dasein in history. Time is not the unending succession of more-or-less identical moments, nor is it the cyclical return of the same, finite, natural constellations (as in Nietzsche's eternal recurrence of the same[53]). Rather, Heidegger seems to be pointing to the essential duality of history, which always throws up choices, alternatives, conflicts—although not in the way we are accustomed to thinking of them. History mostly does *not* confront us with choices about which we can "decide." Rather, the de-cision is whether to *have* a history or not—i.e., whether to respond to the call which the tradition holds for us or to ignore it and accelerate our activity, in the belief that speed alone or in combination with brute force will lead to an "end," a *telos*, which can fill the emptiness of that kind of living.

We have seen that a vital part of Heidegger's thought about art, and about *"Wie wenn am Feiertage..."* in particular, is involved with the issue of the "decision." In the earlier essays, Heidegger effectively destroyed the structures of art and history, as they are commonly formulated. He set the literary artwork "free" to let the truth go to work. There is, however, a danger in this liberating destruction of inappropriate structures. The danger inherent in attacking the metaphysical system of stable, "objective" time is that man thereby is freed for nothing, for a nihilism without distinctions of value—or rather, with only meaningless, arbitrary, or heuristic values. Heidegger claimed, of course, that the view of art and history fostered by inappropriate scientific-technical world-views created precisely that valueless, *gleichgültig*, pseudo-freedom. Only when the artwork is brought back to rest on its foundation in the de-cision of an historical people can it "go to work." How does this happen? In §45 of the *Beiträge zur Philosophie* he offers an answer.

> Die Entscheidung *muß jenen Zeit-Raum*, die Stätte für die wesentlichen Augenblicke schaffen, in der der höchste Ernst der Besinnung in eins mit der größten Freudigkeit der Sendung zu einem Willen des Gründens und Bauens aufwächst, dem auch keine Wirrnis fern bleibt. [BP,98]

What Heidegger seems to be describing here is clearly very close to the *Feiertag* as we have seen it above. He sees the decision as opening up the time-space for those "essential moments" in which a people come together in the ground of their Being and which combine "the greatest seriousness" with "the greatest joyfulness."[54] Normally, the fundamental decisions of our society and of our history, the origins and the fate of our "world," are hidden. On the *Feiertag,* the poet steps into the clearing and his words call the Holy and the people together. He establishes the time of his people and sets in motion a "new beginning." The *Feiertag* discloses the kind of time in which the people assemble and reconsecrate themselves to the basis of their existence and identity. It is a time of renewal after the storms of the *Alltag*. Of course, there is also great danger in such moments; without a counterbalancing seriousness, the freedom from the *Alltag* could lead to the total failure to grasp the decisions of the community in their original light and result in simply more and more meaningless "culture." The *Feiertag* requires the advent of the Holy. Without it, it becomes just one more day and loses its special restorative power.

It seems clear that Heidegger's thoughtful engagement with Hölderlin was more than merely an exploitation of a dead poet's verses for the propagation of his own brand of mysticism. Hölderlin's *Dichten* entered into and shaped Heidegger's thought fundamentally, as exemplified by the relationship of the *Feiertag* which Heidegger "sounded" (*erläutert*) in his reading of *"Wie wenn am Feiertage..."* and the decision he saw facing us in the modern age. The dialogue which was started up between them in the essays of the 1930's has proven to be less one-sided than it might at first have seemed. Moreover, Hölderlin's words themselves have taken on new meaning through this encounter with Heidegger. They have brought a kind of temporality to light which escapes the narrow philological approach with which this investigation began. They bring into focus the importance of understanding time and history in the sense of a primordial "time-space" where the decision about our future can be made. *"Wie wenn am Feiertage..."* does not announce a new "time," as is so often asserted. It rather brings us before the de-cisions which have constituted the time of our own historical age. It dares to hear the calling of the Holy, which calls from the most original origin of our historical humanity, and to affirm that calling. It itself thus becomes an origin: *"eine ausgezeichnete Weise, wie die Wahrheit seiend, d.h. geschichtlich wird."* [UK,64]

Our discussion of the *Feiertag* took us into history. *"Wie wenn am Feiertage..."* discovered history to be arriving NOW, in the present. Can we too discover the *Feiertag*?

Notes

Chapter One: The Problematics of Literary Temporality

1. Cf. Emil Staiger, *Die Zeit als Einbildungskraft des Dichters* (Zürich and Leipzig: Max Niehans, 1939) 66ff.

2. "Hölderlins Dichtung im Zeitalter des Idealismus," in *Hölderlin-Aufsätze* (Frankfurt a. M.: Insel, 1972) 29.

3. Richard Unger, *Hölderlin's Major Poetry: The Dialectics of Unity* (Bloomington: Indiana U Press, 1972) 119.

4. Alessandro Pellegrini, *Friedrich Hölderlin: Sein Bild in der Forschung* (Berlin: Walter de Gruyter, 1965) 18.

5. First published in the *Verlag der Blätter für die Kunst*, then in 1913 as first volume of *Hölderlin: Sämtliche Werke* (Berlin: Propyläen). This information is from Albrecht Seifert, *Untersuchungen zu Hölderlins Pindar-Rezeption* (Munich: Wilhelm Fink, 1982) 16.

6. *Hölderlin-Studien* (Frankfurt a.M.: Insel, 1967).

7. Pellegrini, 55.

8. *'Hölderlin und das Wesen der Dichtung,'* in *Erläuterungen zu Hölderlins Dichtung* 5th. ed. (Frankfurt a.M.: Klostermann, 1981) 34.

9. "On the Art of Poetry," in T. S. Dorsch, ed. and trans., *Classical Literary Criticism* (London: Penguin Books, 1965) 31.

10. *The Deconstruction of Time* (Atlantic Highlands, NJ: Humanities P Intl., 1989) 300.

11. "Heidegger and Hölderlin: The Limits of Language," *The Personalist* 44 (1963) 5.

12. *Heidegger, Thought, and Historicity* (Ithaca: Cornell U Press, 1986) 175.

13. "Poetry and Language in Heidegger," in Joseph J. Kockelmans, ed., *On Heidegger and Language* (Evanston, IL: Northwestern U Press, 1972) 66.

14. *Erläuterungen*, Preface to second edition, 7.

15. *Being and Time*, 62.

16. *Heidegger's Language and Thinking* (Atlantic Highlands, NJ: Humanities P Int'l., 1985) x.

17. HWD,44.

Chapter Two: Hymn or Warning?

1. Seifert (104–17) gives the most thorough description of the state of this debate, although I do not agree with his reasons for discarding Beißner's dating of late 1799. In my opinion, however, any more precise dating than I have given above remains speculation.

2. Cf. *Hölderlin: Sämtliche Werke* (called the *"Grosse Stuttgarter Ausgabe"*), ed. Friedrich Beißner, 4th. ed. (Stuttgart: W. Kohlhammer, 1951) vol. II/2, 667.

3. Reproduced above, pp. viii–ix.

4. Hölderlin's idiosyncratic spelling will be used throughout this work to indicate the peculiar temporal significance of this common word.

5. Throughout this study I will employ roman numerals to indicate strophes (stanzas), followed by arabic numerals to indicate verses within each strophe. This seems more reasonable to me than the standard practice of continuous numbering, since it allows the structure of the poem, which is carefully divided into equal strophes and groups of strophes (triads) to be kept in view.

6. Günther Wohlfahrt, *Der Augenblick* (Freiburg, Munich: Karl Alber, 1982) 70 et passim.

7. Cf. Rolf Peter Norgaard, "Placing Voice: Story and Speech in the Poetry of Wordsworth and Hölderlin," diss., Stanford, 1982, 302.

8. For example, Momme Mommsen, "Die Problematik des Priestertums bei Hölderlin," *Hölderlin-Jahrbuch* 15 (1967/68): 60.

9. "Der andere Pfeil," *Hölderlin-Studien* (Frankfurt a.M.: Insel, 1967) 33–54; expanded version also in P. S.: *Einführung in die literarische Hermeneutik*, ed. Jean Bollack and Helen Stierlin (Frankfurt a.M.: Suhrkamp, 1975) 217–323.

10. "Hölderlins Hymne *'Wie Wenn am Feiertage...'*" in: *Erläuterungen zu Hölderlins Dichtung* (1941, originally 1939).

11. "Hölderlins erste Hymne," *DVLG* 17 (1939):221–51.

12. "Das ewige Herz in Hölderlins Dichtung," in *Worte und Werte*: *Bruno Markwardt zum 60. Geburtstag*, eds. Gustav Erdmann and Alfons Eichstaedt (Berlin: de Gruyter, 1961) 297.

13. *Hölderlin. Sämtliche Werke*, historisch-kritische Ausgabe, unter Mitarbeit von Friedrich Seebaß besorgt durch Norbert von Hellingrath, 3rd. ed., vol. 4, (Berlin: Propyläen, 1913).

14. "Hymne," 220.

15. I take this felicitous expression from William A. O'Brien's article, "Getting Blasted: Hölderlin's 'Wie Wenn am Feiertage...'," *MLN* 94(1) (1979): 569–586.

16. Wilhelm und Jacob Grimm, eds., *Deutsches Wörterbuch* (1852–1984), vol.12, 90.

17. Hermann von Fischer ed., *Schwäbisches Wörterbuch* (Tübingen: Laupp, 1914) vol. III.

18. Some commentators still identify it rather too casually in this way. E.g., Seifert, 158. Cf. Grimm, *Wörterbuch*, vol.3.

19. *"Das Eine in sich selber unterschiedne."* [sic] (Differentiation-within-unity) Quoted in *Hyperion*, GSA, III, 81.

20. *Der Gang des Andenkens* (Stuttgart: Klett-Cotta, 1987) 20.

21. *Words after Speech: A Comparative Study of Romanticism and Symbolism* (London: MacMillan, 1986) 50.

22. Seifert, 122f.

23. "Hölderlins idealistischer Dichtungsbegriff in der poetologischen Tradition des 18. Jahrhunderts," *Hölderlin-Jahrbuch* 22 (1980/81): 116.

24. *Text, Geschichte und Subjektivität in Hölderlins Dichtung: Uneßbarer Schrift gleich*, 1985, 181.

25. Seifert, 126.

26. I keep the feminine pronoun when referring to "Nature" to reflect its gender in German. The significance of its gender becomes apparent later in the poem.

27. *Hölderlin's Major Poetry*, 111.

28. Seifert also points to the externality of this description (*Untersuchungen*, 185).

29. For a representative sample of the literature concerning this point, see Paul DeMan: "Heidegger's Exegeses of Hölderlin," in: *Blindness and Insight*, 2nd ed., rev. (Minneapolis: U of Minnesota Press, 1971) 246–66; Günther Wohlfahrt: *Der Punkt. Ästhetische Meditationen* (Freiburg, Munich: Karl Adler, 1986) 180; Maurice Blanchot: "La Parole Sacrée de Hölderlin," in: M.B., *La part du feu* (Paris: Gallimard, 1949) 137; and Hans-Jost Frey: "Das Heilige und das Wort," in: H.-J.F., *Studien über das Reden der Dichter* (Munich: Wilhelm Fink, 1986).

30. There is a parallel here to Hölderlin's roughly contemporaneous essay, "Das Werden im Vergehen": *"Denn die Welt aller Welten, das Alles in Allen, welches immer ist, stellt sich nur in aller Zeit ... dar ...."* (italics in original) GSA, IV/1, 282.

31. A partial listing of those representing this view would have to include: Wilhelm Michel, *Friedrich Hölderlin* (Weimar: Lichtenstein, 1925) 86–7; Ernst Ludwig Stahl, "Hölderlin's Poetic Mission," *GLL* 2 (1948/49) 54; Peter Szondi, "Der andere Pfeil," in *Hölderlin-Studien*, 42–3; Walter Müller-Seidel in "Hölderlins Ode *Dichterberuf*," in: W. M.-L., *Die Geschichtlichkeit der deutschen Klassik* (Stuttgart: Metzler, 1983), as well as many others.

32. Unger, 110.

33. Peter Szondi, "Der andere Pfeil," 42.

34. Cf. Herbert Lederer, *Reference Grammar of the German Language* (New York: Charles Scribner's Sons, 1969) §F 421.

35. Seifert, 238.

36. "Chaos," *Griechisches etymologisches Wörterbuch*, Hjalmar Frisk, ed. (Heidelberg: Carl Winter, 1970).

37. *Theogony*, 116. Edited and with Commentary by M. L. West (Oxford: Clarendon Press, 1966). The editor translates it as "Chasm," i.e., a "yawning space," and says it is "dark and gloomy." He notes there is some disagreement as to its meaning in Hesiod's time (pp. 192-3).

38. *Metamorphoseon*, I,7 (*"rudis indigestaque moles"*).

39. Alfred Doppler, *Der Abgrund* (Graz: H. Bohlaus Nachf., 1968) 89.

40. Arthur Häny, *Hölderlins Titanenmythos* (Zürich: Atlantis, 1948) 18.

41. "Äther und Abgrund in Hölderlins Dichtung," *Frankfurt aber ist der Nabel dieser Welt*, eds. Christoph Jamme and Otto Pöggeler (Stuttgart: Klett-Cotta, 1983) 355.

42. Not to be confused with Heidegger's sense.

43. Unger, 112.

44. Seifert, 187.

45. Cf. Nägele, 187, who describes this as *"vorsprachliche Erkenntnis."* As long as language is taken solely as words (*paroles*), this may be true. It would seem,

though, that even non-verbal acquaintance requires language (else what is "known"?). This is an area of fundamental disagreement between an empirical and a Heideggerian view of language.

46. *Hölderlins später Widerruf in den Oden "Chiron," "Blödigkeit" und "Ganymed"* (Tübingen: Max Niemeyer, 1978) 166.

47. *Widerruf*, 170.

48. *"Prometheus."*

49. A. O'Brien calls it "ominous." ("Getting Blasted," 576)

50. For an overview and discussion, see W. A. O'Brien, "Getting Blasted."

51. Norgaard (286) shows the effect of syntax in framing the moment of divine revelation.

52. I would therefore have to take exception to Nägele's somewhat casual psychologizing.(*Text, Geschichte*, 188f).

53. Lachmann, 223.

54. Alice Kuzniar, *Delayed Endings: Nonclosure in Novalis and Hölderlin* (Athens, GA/London: U. of Georgia Press, 1987) 158. She describes it as the poet following "the logic of his metaphor."

55. *Problematik*, 58.

56. Peters, *Greek Philosophical Terms* (New York: New York University Press, 1967) 154f.

57. "Der andere Pfeil." The entire essay deals with this question. See for rebuttal W. A. O'Brien, "Getting Blasted."

58. *Problematik*, 60.

59. *Problematik*, 60–68.

60. See discussion in O'Brien, "Getting Blasted."

61. *Untersuchungen*, 310–11.

62. *Placing Voice*, 292.

63. Only Momme Mommsen (*Problematik*, 67) does not seem to share this view. He asserts that this is a collective, not a private, fate.

64. Cf. A. Kuzniar, 160.

65. *Placing Voice*, 302.

66. "Hölderlin: Philosophie, Pantheismus, Temporalität," *Orbis Literarum* 28(1) (1973): 29.

67. Cf. "Das Werden im Vergehen," GSA IV/1, 282–87. In the *Frankfurter Ausgabe*: *Friedrich Hölderlin, Sämtliche Werke*, ed. Dietrich Sattler (Frankfurt am Main: Roter Stern, 1975–), it is in vol. XIV, 135–173, where it is titled "das untergehende Vaterland...".

Chapter Three: How to Find an Artwork

1. "Sprachkunstwerk." *Deutsches Wörterbuch,* Gerhardt Wahrig, ed. (Gütersloh, Munich: Bertelsmann Lexikon Verlag, 1986).

2. *Heidegger's Language and Thinking*, 2.

3. Martin Heidegger, *Holzwege* (Frankfurt am Main: Klostermann, 1950, 1980). Cf. Joseph Kockelmanns, *Heidegger on Art and Artworks* (Dordrecht: Martinus Nijhoff, 1985).

4. Cf. Kockelmanns, 90.

5. *Heidegger's Estrangements* (New Haven: Yale U P, 1989) xviii.

6. When the German text seems somewhat obscure or capable of different interpretations, I will include an English translation to avoid misunderstanding. The authorized English translation is that of Albert Hofstadter in *Poetry, Language, Thought* (New York: Harper Colophon, 1971), abbreviated here as PLT. All other translations are my own.

7. Kockelmanns, 140.

8. *Heidegger's Estrangements: Language, Truth, and Poetry in the later Writings* (New Haven: Yale U Press, 1989) 30.

9. Cf. especially Charles H. Kahn, *The Art and Thought of Heraclitus* (Cambridge: University Press, 1979, 1983) 202–209. The literature on Heidegger's relationship with Heraclitus is too extensive to summarize here, although little has to date been written on the similarity of their *ways of thinking*. Suffice it to say that much in Heidegger's own "phenomenological hermeneutic" derives from his dialogue with the Greek philosopher.

10. "Dem Dichter Vor-Denken. Aspekte von Heideggers 'Zwiesprache' mit Hölderlin im Kontext seiner Kunstphilosophie," *Zeitschrift für philosophische Forschung* 38/2 (Ap–Ju 1984): 195.

11. PLT, 77. In that context it is used with respect to history (*Entrückung*).

12. Bruns, 44.

13. PLT, 67. The whole sentence reads: *"Knowing that remains a willing, and willing that remains a knowing, is the existing human being's entrance into and compliance with the unconcealedness of Being."*

14. Kockelmanns, 181.

15. A "having-been-decided." In English from the Latin *de-cidere*, literally "to cut open." The German word is an analogous construction.

16. "dichten," *Deutsches Wörterbuch*, vol. 2, 1057.

17. B&T, 37.

18. See also HWD, 46, in which Heidegger refers to sagas (*die Sagen*) in Hölderlin's "Stimme des Volks" as *"a people's way of remembering its belongingness to the totality of beings" (zum Seienden im Ganzen).*

19. *The Question of Language in Heidegger's History of Being* (Atlantic Highlands, NJ: Humanities Press, 1985) 36.

20. Or "transporting," as Albert Hofstadter has it (PLT, 77).

21. After the publication of *"Der Ursprung des Kunstwerkes,"* Heidegger added two supplemental sections, a postscript and "addenda." Since they do not

significantly affect the course of this reading of the text, they will not be discussed.

Chapter Four: Poetry and the Decision

1. *Heidegger's Language and Thinking*, 4.

2. See above, pp. 11–12.

3. Max Kommerell, letter to Heidegger of 29. 7. 1942, reproduced in: M.K.: *Briefe und Aufzeichnungen 1919–1944*, ed. Inge Jens Olten (Freiburg i. B.: Walter, 1967) 400–1.

4. The most notable critics include Theodor Adorno, "Parataxis," *Noten zur Literatur*, ed. Rolf Tiedemann (Frankfurt a. M.: Suhrkamp Veralg, 1974) 452ff.; and Paul de Man, "Heidegger's Exegeses of Hölderlin," *Blindness and Insight: Essays in the Rhetoric of Contemporary Criticism*, 2nd ed., (Minneapolis, MN: U of Minnesota Press, 1983) 246–266.

5. Karsten Harries, "Heidegger and Hölderlin: The Limits of Language," *The Personalist*, 44 (1963): 5 –23.

6. "essence," *A Dictionary of Philosophy*, ed. Antony Flew (New York: St. Martin's, 1979).

7. *"Umwertung politisch profilierter Autoren wie Hölderlin, Kleist und Büchner zu präfaschistischen Repräsentanten des 'heldischen Pessimismus' gehört zu den traurigsten Kapiteln der deutschen Literaturgeschichte."* Beutin, et. al., *Deutsche Literaturgeschichte von den Anfängen bis zur Gegenwart* (Stuttgart: Metzler, 1979) 307-8.

8. Through *Versagen* and *Verstellen* (cf. UK, 39).

9. "zeugen," *Deutsches Wörterbuch*, ed. Gerhardt Wahrig et. al, 1463.

10. SuZ, 42.

11. B&T, 151.

12. *Being-in-the-World: A Commentary on Heidegger's Being and Time, Division I* (Cambridge, MA: MIT Press, 1991) 23.

13. Heidegger often uses the expression, *"die Entscheidung fällt (aus)"* (literally, the decision "falls out," or "turns out"), which avoids the temptation of seeing it as an "activity" or "choice." The English expression, "to make a decision," unfortunately suggests just such a psychological act.

14. SuZ, 326. Emphasis in original.

15. With thanks to H. Dreyfus for calling attention to this passage.

16. Theodore Kisiel, "The Hermeneutics of Gadamer and Heidegger," *Hermeneutics and Praxis*, ed. Robert Hollinger (Notre Dame, Ind: University of Notre Dame Press, 1985) 9–10.

17. Dreyfus, 24.

18. *Heidegger's Estrangements*, xviii. Cf. also 38.

19. UK, 61. See also discussion above, 89–90.

20. We can clearly see here, in my opinion, another allusion to the position of the poet in *"Wie wenn am Feiertage...,"* in which the poet insists that poets are expected to bear the god's lightning bolt *"dem Volk ins Lied gehüllt."* (St. VII,6). Although Heidegger restricts himself to only one explicit mention of our poem in this essay, the frequency of his allusions to it shows how near to his thoughts it was.

21. *"The 'essence' of this entity lies in its 'to-be' [Zu-sein]."* B&T, 67.

22. *The Question of Language,* 38.

23. For Nietzsche, see Walter Kaufmann, *Nietzsche: Philosopher, Psychologist, Antichrist* (Cleveland and New York: Meridian, 1950, 1964) 19f. For Hölderlin, cf. the reactions of Schiller and Goethe, cited in Pellegrini, 13–19ff, as well as Uwe Henrik Peters, *Hölderlin: wider die These vom edlen Simulanten*, (Reinbeck bei Hamburg: Rowohlt, 1982) 189.

Chapter Five: Heidegger's Process of Reading Hölderlin

1. "HOLDERLINS HYMNE *'Wie wenn am Feiertage...'*". Actually, the title of the essay as it appears in the *Erläuterungen* is simply *"Wie wenn am Feiertage...,"* but I have retained the title under which the essay originally appeared (Halle

a.d.S.: Max Niemeyer, 1941) to avoid unnecessary confusion with the text of the poem.

2. Cf. Joseph P. Fell, *Heidegger and Sartre: An Essay in Being and Place* (New York: Columbia U P, 1979), 164–184; also Hubert L. Dreyfus, *Being-in-the-World: A Commentary on Heidegger's* Being and Time, *Division I*, (Cambridge, MA: MIT Press, 1991) 90.

3. SuZ, 42, emphasis added.

4. B&T, 195.

5. See *"Der Ursprung des Kunstwerkes,"* 3.

6. Cf. Dominique Janicaud, "Metamorphosis of the Undecidable," *Graduate Faculty Philosophy Journal* 13 (1) (1988): 125–41. Originally published in Dominique Janicaud and Jean-François Mattéi, *La Métaphysique à la limite. Cinque études sur Heidegger* (Paris: Presses Universitaires de France, 1983).

7. B&T, 58.

8. HWD, 34.

9. Curiously, the text of this passage says the opposite: *"Der hier zugrunde gelegte Text beruht, nach den urschriftlichen Entwürfen erneut geprüft, auf dem folgenden Versuch einer Auslegung."* In a letter to Heidegger, Detlev Lüders called his attention to the paradox in this formulation. In his reply (24. February 1953), Heidegger recognized the error immediately (*"Sie haben recht"*) and confirmed that the sentence should read: *"Der folg.* [sic] *Versuch einer Auslegung beruht auf dem nach den handschriftlichen Entwürfen erneut geprüften Text."* Although Heidegger expressed his intention to correct this error in the next edition of the *Erläuterungen* (3. ed.,1963), somehow this never took place. Lüders presumes (and I agree) that this was merely an oversight. He discusses this matter in a *Sonderdruck* of the *Jahrbuch des freien deutschen Hochstifts* (Tübingen: Niemeyer, 1977)247-50.

10. Eduard Lachmann, "Hölderlins erste Hymne," *Deutsche Vierteljahrsschrift* 17 (1939): 221–251. Friedrich Beißner, *Hölderlins Übersetzungen aus dem Griechischen* (Stuttgart: 1933) 96–103. Wilhelm Michel, "Hölderlins Hymne an die Dichter," *Das Reich* 3 (1918–19), Buch 3: 451–53 (also in W. M.: *Friedrich Hölderlin* [Weimar, 1925]:81–91).

11. I. Kings, XIX, 11–12.

12. *Erläuterungen*, 195.

13. This philological *faux pas* has disturbed many of Heidegger's readers, who feel he thereby misses the darkness in Hölderlin's vision. Cf. Christoph Jamme, "Dem Dichter Vor-Denken," 212, Paul de Man, "Heidegger's Exegeses," 261, and Karsten Harries, "Heidegger and Hölderlin," 22, among others.

14. Indeed, as Michael Murray has pointed out, Heidegger is in fact well aware of the poem's "incompleteness" ("Heidegger's Hermeneutic Reading of Hölderlin: The Signs of Time," 48).

15. In this respect, I agree with Hans-Jost Frey, who says, *"Das Wort, welches das Heilige nennt, ist nicht selbst das Heilige."* ("Das Heilige und das Wort," *Studien über das Reden der Dichter*, [Munich: Wilhelm Fink, 1986] 138).

16. Hellingrath's edition, except for Heidegger's replacement of the comma by a period.

17. I have capitalized ontological NOW to distinguish it from historical "now."

18. Cf. Dreyfus, 90.

19. *Erläuterungen*, 195.

20. Paul de Man, "Heidegger's Exegeses," 253 et passim.

21. *Heidegger's Estrangements*, 64.

Chapter Six: Conclusion

1. *Confessions*, Book XI, Ch.14.

2. *The Phenomenology of Internal Time-Consciousness* (Bloomington: Indiana U. Press, 1964) 50–76.

3. Ovid, *Metamorphoseon*, Book X, 518.

4. UK, 17.

5. UK, 64; cf. also above, Chapter Three, 112–13.

6. B&T, 92.

7. Cf. G. Bruns, *"Heidegger, however, doesn't have a method."* (*Heidegger's Estrangements*, 64). Also Richard E. Palmer's critique of interpretive methods in *Hermeneutics*, 247–50.

8. *Martin Heidegger, The Way and the Vision*, (Honolulu: U Press of Hawaii, 1976) 46.

9. Mehta, 46.

10. Mehta, 47.

11. Heidegger, of course, was not the first reader to emphasize the importance of Nature. Cf. Eduard Lachmann: *"Das Wort von der mächtigen, göttlich schönen Natur zieht alle Worte in sein Geheimnis."* in "Hölderlins erste Hymne," *DVjG* 17 (1939): 221–51.

12. HH, 51–2.

13. "Heidegger's Exegeses of Hölderlin," in P. de M.: *Blindness and Insight* (Minneapolis: U of Minnesota P, 1971, 1983) 250, 253.

14. *"The coming of the glorified Christ as the climax of salvation history."*—"Parousia," in *New Catholic Encyclopedia*, (New York: McGraw-Hill, 1967–79). Heidegger himself distinguishes the Greek understanding from his own, in *Einführung in die Metaphysik* (Frankfurt a.M.: Klostermann, 1983) 65.

15. "Heidegger's Exegeses," 252.

16. Two important and frequently cited studies are by Hinrich Knittermeyer, "Fest und Feier: Ein Beitrag zum Wortgebrauch Hölderlins," *Hölderlin-Jahrbuch* (1950) 47–52, and Rudolf Nießen, "Das Phänomen der Feier in der Dichtung Hölderlins: eine religionsphilosophische Untersuchung," diss., Munich, 1954.

17. Seifert, 158.

18. *Friedensfeier: Eine Einführung in Hölderlins Christushymnen* (Munich: C.H. Beck, 1959) 67.

19. Knittermeyer, for example, emphasizes that Hölderlin did not primarily have the Christian "holiday" in mind: *"Es ist die Heiterkeit und offene Freude des griechischen Festes, die den Dichter* [i.e., Hölderlin] *beglückt."* ("Fest," 52)

20. *Works and Days*, 780f.

21. For a discussion, see G. E. R. Lloyd, "Views on Time in Greek Thought," *Cultures and Time*, ed. Paul Ricoeur (Paris: Unesco Press, 1976) 117–148.

22. G. Liddell, R. S. Scott, eds., *Greek-English Lexicon* 9th ed. (London: Oxford U Press, 1843–1948) vol. 1, 859.

23. John R. Wilson, "KAIROS as 'Due Measure,'" *Glotta* 58(3) (1980) 177–204. Also: Hjalmar Frisk, ed., *Griechisches etymologisches Wörterbuch* (Heidelberg: Carl Winter, 1960) 755: *"rechtes Maß."*

24. L. West, "Excursus II: Time Reckoning," in: Hesiod, *Works and Days* (Oxford: Clarendon Press, 1978) 377–8.

25. Nießen, *Das Phänomen der Feier*, 40.

26. West, 378.

27. Nießen, 40.

28. "Dichtung und Zeit in Hölderlins Werk," unpublished dissertation, Tübingen, 1955: 239.

29. The dual nature of the *Feier* is preserved in the term *feierlich* (solemn), and *Todesfeier*, on the one hand; as well as *feiern*, which today usually refers to partying, celebrating, etc.: *" etw. festlich begehen"* (Wahrig). In English, this distinction appears as "holiday" (time off, time for recreation), which manifests its derivation from "holy day" (cf. OED, "holiday").

30. "Zur Funktion der Feste in Athen im 5. Jahrhundert vor Christus," *Das Fest*, eds. Walter Haug and Rainer Warning, (Munich: Wilhelm Fink, 1989) 589–90.

31. *Greek Religion*, transl. John Raffan (Cambridge, MA: Harvard U Press, 1985) 103.

32. *"Tage, an denen keine Geschäfte vorgenommen werden,"* "Feier," *Deutsches Wörterbuch*, vol.3.

33. Cf. Nägele, *Text, Geschichte*, 182. I cannot agree with him, however, that the *Landmann* is also the "poetic subject" whose inner split causes the poem's final fragmentation.

34. He, of course, avoids most of the difficulty by simply ignoring the last two strophes, as was noted in Chapters Two and Five.

35. In this, it resembles Heidegger's description of Nature: *"...das Sichöffnen, das aufgehend zugleich zurückgeht in den Hervorgang . . ."* [HH 56, also p. 161 above]

36. In Greece, festivals were rarely interrupted, even during war. Cf. Meier 569, in Haug, *Das Fest.*

37. *Greek Religion*, 99.

38. *Greek Religion*, 255.

39. As Burkert notes, *"...sacred* [hieros] *is the day on which the gods are present..."* (269)

40. "Das Phänomen der Feier," 38.

41. Principally in §§ 23–27 of M.H., *Hölderlins Hymne "Andenken,"* (Frankfurt a. M.: Vittorio Klostermann, 1982), ed. Curd Ochwadt. Here abbreviated "A" to avoid confusion with the lecture on *"Wie wenn am Feiertage...."*

42. A, 60.

43. B. Benn, *Hölderlin and Pindar* ('S-Gravenhage: Mouton, 1962) 155–6.

44. I owe this information about contemporary French festivals to a talk given by Karlheinz Stierle at the conference "Hegel and Hölderlin" held at Yale University, 9.–11. October 1987. His paper has subsequently been published as "Die Friedensfeier: Sprache und Fest im revolutionären und nachrevolutionären Frankreich und bei Hölderlin," in: *Das Fest*. Eds.

Walter Haug and Rainer Warning (Munich: Wilhelm Fink, 1989) 481–525.

45. As Stierle notes about Hölderlin's other great "*Feier*" poem: "*Hölderlins Hymne ist nicht Teil einer Friedensfeier, sie ist die Friedensfeier selbst.*" "Die Friedensfeier," Haug, *Das Fest* 499.

46. Cf. HWD,40.

47. August von Herrmann, ed., (Frankfurt a.M.: Klostermann, 1989), vol. 65 of the *Gesamtausgabe*.

48. UK,64. Cf. Chapt. 2, above.

49. Cf. SuZ,127: "*Das Man* entlastet *so das jeweilige Dasein in seiner Alltäglichkeit.*" (Heidegger's emphasis)

50. *Being-in-the-World,* 151–2. There is no commonly-accepted translation of this important term. Macquerie and Robinson use "the They." (§27) Dreyfus argues, and I concur, that this misses the "appeal to normativity" which is crucial to Heidegger's use of the term.

51. For a discussion of *Wissen* and *Wollen*, see UK, 53–4 and above, 107–8.

52. Cf. Friedrich Nietzsche, "*Zu allem Handeln gehört Vergessen...*" ("Vom Nutzen und Nachteil der Historie für das Leben," *Unzeitgemäße Betrachtungen,* Frankfurt a. M.: Insel,1981) 99.

53. Nietzsche first expounded in *Fröhliche Wissenschaft* (1882) §341:"*Wie, wenn dir eines Tages oder Nachts ein Dämon in deine einsamste Einsamkeit nachschliche und dir sagte: 'Dieses Leben, wie du es jetzt lebst und gelebt hast, wirst du noch einmal und noch unzählige Male leben müssen...'* "

54. Cf. above, ftn. 30.

Bibliography

Adorno, Theodor W. *Noten zur Literatur.* Ed. Rolf Tiedemann. Frankfurt am Main: Suhrkamp Verlag, 1974.

Aristotle. "On the Art of Poetry," in *Classical Literary Criticism*. Ed. T. S. Dorsch. Harmondsworth, Middlesex: Penguin, 1965.

Beißner, Friedrich. *Hölderlins Übersetzungen aus dem Griechischen.* 1933. 2nd ed. Stuttgart: Metzler, 1961.

Bernasconi, Robert. *The Question of Language in Heidegger's History of Being.* Atlantic Highlands, NJ: Humanities Press, 1985.

Benn, M[aurice] B. *Hölderlin and Pindar.* 'S-Gravenhage: Mouton, 1962.

Beutin, Wolfgang, and Klaus Ehlert, Wolfgang Emmerich, Helmut Hoffacker, Bodo Lecke, Bernd Lutz, Ralf Schnell, Peter Stein, Inge Stephan. *Deutsche Literaturgeschichte von den Anfängen bis zur Gegenwart* Stuttgart: J. B. Metzler, 1979.

Biemel, Walter. "Poetry and Language in Hölderlin and Heidegger." *On Heidegger and Language*. Ed. and trans. Joseph J. Kockelmans. Evanston, IL: Northwestern UP, 1972.

Binder, Wolfgang. "Äther und Abgrund in Hölderlins Dichtung." *Frankfurt aber ist der Nabel dieser Welt*. Eds. Christoph Jamme and Otto Pöggeler. Stuttgart: Klett-Cotta, 1983.

———. *Hölderlin-Aufsätze*. Frankfurt am Main: Insel, 1970.

———. "Dichtung und Zeit in Hölderlins Werk." Diss., Tübingen, 1955.

Blanchot, Maurice. "La parole 'sacrée' de Hölderlin." *La Part du Feu.* Paris: Gallimard, 1949.

Bruns, Gerald. *Heidegger's Estrangements: Language, Truth, and Poetry in the Later Writings*. New Haven, CT: Yale UP, 1989.

Burkert, Walter. *Greek Religion*. Trans. John Raffan. Cambridge, MA: Harvard UP, 1985.

"Chaos." *Griechisches etymologisches Wörterbuch*. Ed. Hjalmar Frisk. Heidelberg: Carl Winter, 1970.

Coates, Paul. *Words After Speech: A Comparative Study of Romanticism and Symbolism.* London: MacMillan, 1986.

Doppler, Alfred. *Der Abgrund*. Graz: H. Bohlaus Nachf., 1968.

Dreyfus, Hubert. *Being-in-the-World: A Commentary on Heidegger's* Being and Time, *Division I.* Cambridge, MA: MIT Press, 1991.

Fell, Joseph P. *Heidegger and Sartre: An Essay in Being and Place*. New York: Columbia U P, 1979.

Fischer, Herrmann von, ed. *Schwäbisches Wörterbuch*. Tübingen: Laupp, 1914.

Flew, Antony. *A Dictionary of Philosophy*. New York: St. Martin's, 1979.

Frey, Hans-Jost. *Studien über das Reden der Dichter.* Munich: Wilhelm Fink, 1986.

Fynsk, Christopher. *Heidegger: Thought and Historicity*. Ithaca and London: Cornell UP, 1986.

Grimm, Wilhelm and Jakob, eds. *Deutsches Wörterbuch*. 1860. Rpt. Munich: Deutscher Taschenbuch Verlag, 1984.

Häny, Arthur. *Hölderlins Titanenmythos.* Zürich: Atlantis, 1948.

Harries, Karsten. "Heidegger and Hölderlin: The Limits of Language." *The Personalist* 44 (1963): 5–23.

Haug, Walter and Rainer Warning, eds. *Das Fest.* Munich: Wilhelm Fink, 1989.

Häussermann, Ulrich. *Friedensfeier: Eine Einführung in Hölderlins Christushymnen.* Munich: C. H. Beck, 1959.

Heidegger, Martin. *Beiträge zur Philosophie (Vom Ereignis).* Frankfurt am Main: Vittorio Klostermann, 1989.

———. *Einführung in die Metaphysik,* 1953. Tübingen: Max Niemeyer, 1983.

———. *Erläuterungen zu Hölderlins Dichtung.* 1944. 5th. ed. Frankfurt am Main: Vittorio Klostermann, 1971.

———. HÖLDERLINS HYMNE *"Andenken."* Frankfurt a. M.: Vittorio Klostermann, 1982.

———. HÖLDERLINS HYMNE *"Wie wenn am Feiertage."* Halle a.d.S.: Max Niemeyer, 1939. Included as part of *Erläuterungen*.

———. *Holzwege*. 1959. 6th ed. Frankfurt am Main: Vittorio Klostermann, 1980. Includes *"Der Ursprung des Kunstwerkes"* ("The Origin of the Work of Art"). Trans. Albert Hoftstadter. *Poetry, Language, Thought*. New York: Harper & Row, 1971.

———. *Sein und Zeit*. 1927. 16th ed. Tübingen: Max Niemeyer, 1986. Trans. John Macquarrie and Edward Robinson. *Being and Time.* New York: Harper & Row, 1962.

Henrich, Dieter. *Der Gang des Andenkens.* Stuttgart: Klett-Cotta, 1986.

Hesiod. *Works and Days.* Ed. and commentary M. L. West. Oxford: Clarendon Press, 1978.

———. *Theogony*. Ed. and commentary M. L. West. Oxford: Clarendon Press, 1966.

Hölderlin, Friedrich. *Sämtliche Werke.* 1913. Vol. IV. Begun by Norbert von Hellingrath, completed by Friedrich Seebaß and Ludwig Pigenot. 3rd ed. Berlin: Im Propyläen Verlag, 1943.

———. *Sämtliche Werke.* Vols. II and IV (Stuttgarter Hölderlin-Ausgabe, "Große Stuttgarter Ausgabe"). Ed. Friedrich Beißner. Stuttgart: Kohlhammer, 1951.

———. *Sämtliche Werke. Frankfurter Ausgabe.* Ed. Dietrich Sattler. Frankfurt am Main: Roter Stern, 1975–).

Husserl, Edmund. *The Phenomenology of Internal Time-Consciousness.* Ed. Martin Heidegger. Trans. James S. Churchill. Introduction by Calvin O. Schrag. Bloomington, IN: Indiana U P, 1964.

Jamme, Christoph. "Dem Dichter Vor-Denken. Aspekte von Heideggers 'Zwiesprache' mit Hölderlin im Kontext seiner Kunstphilosophie." *Zeitschrift für philosophische Forschung.* 38(2) (April/June 1984): 191–218.

Janicaud, Dominique. "Metamorphosis of the Undecidable." *Graduate Faculty Philosophy Jounal* 13(1) (1988): 125 –141.

Kahn, Charles H. *The Art and Thought of Heraclitus.* Cambridge (UK): Cambridge U P, 1979, 1983.

Kaufmann, Walter. *Nietzsche: Philosopher, Psychologist, Antichrist.* 1950. New York: Meridian Books, 1964.

Kisiel, Theodore. "The Hermeneutics of Gadamer and Heidegger." *Hermeneutics and Praxis.* Ed. Robert Hollinger. Notre Dame, IN: U of Notre Dame Press, 1985.

Knittermeyer, Hinrich. "Fest und Feier: ein Beitrag zum Wortgebrauch Hölderlins." *Hölderlin-Jahrbuch* (1950): 47–71.

Kockelmans, Joseph J. *Heidegger on Art and Art Works.* Dordrecht: Martinus Nijhoff, 1985.

Kommerell, Max. *Briefe und Aufzeichnungen 1919–1944.* Ed. Inge Jens Olten. Freiburg im Breisgau: Walter, 1967.

Kuzniar, Alice. *Delayed Endings: Nonclosure in Novalis and Hölderlin.* Athens, GA: U of Georgia Press, 1987.

Lachmann, Eduard. "Hölderlins erste Hymne." *DVjG* 17 (1939): 221–51.

Lederer, Herbert. *Reference Grammar of the German Language.* New York: Charles Scribner's Sons, 1969.

Liddell, H. G. and Scott, R. S., eds. *A Greek-English Lexikon*, 1843. 9th ed. Oxford: Clarendon Press, 1948.

Lloyd, G. E. R. "Views on Time in Greek Thought." *Cultures and Time.* Ed. and Intro. Paul Ricoeur. Paris: Unesco Press, 1976.

Lüders, Detlev. "Ein Brief Martin Heideggers zu seinen Hölderlin-Erläuterungen." *Jahrbuch des freien deutschen Hochstifts (Sonderdruck)*. Tübingen: Max Niemeyer, 1977.

de Man, Paul. "Heidegger's Exegeses of Hölderlin." *Blindness and Insight: Essays in the Rhetoric of Contemporary Criticism.* 1971. 2nd. rev. ed. Trans. and Intro. Wlad Godzich. Minneapolis, MN: U of Minnesota Press, 1983.

Mehta, J. L. *Martin Heidegger: The Way and the Vision.* Honolulu: U P of Hawaii, 1976.

Meier, Christian. "Zur Funktion der Feste in Athen im 5. Jahrhundert vor Christus." Haug, 569–591.

Michel, Wilhelm. *Friedrich Hölderlin.* Weimar: Lichtenstein, 1925.

Mommsen, Momme. "Problematik des Priestertums bei Hölderlin." *Hölderlin-Jahrbuch* 15 (1967/68): 53–74.

Mugerauer, Robert. *Heidegger's Language and Thinking.* Atlantic Highlands, NJ: Humanities P Int'l., 1985.

Müller-Seidel, Walter. *Die Geschichtlichkeit der deutschen Klassik.* Stuttgart: J. B. Metzler Verlag, 1983.

Nägele, Rainer. *Text, Geschichte und Subjektivität in Hölderlins Dichtung—'uneßbarer Schrift gleich.'* Stuttgart: J. B. Metzler, 1985.

Nietzsche, Friedrich. *Die fröhliche Wissenschaft.* Munich: Wilhelm Goldmann, 1959.

———. *Unzeitgemäße Betrachtungen.* Frankfurt am Main: Insel, 1981.

"Parousia," *New Catholic Encyclopedia.* New York: McGraw-Hill, 1967–79.

Norgaard, Rolf Peter. "Placing Voice: Story and Speech in the Poetry of Wordsworth and Hölderlin." Diss. Stanford U, 1982.

Nießen, Rudolf. "Das Phänomen der Feier in der Dichtung Hölderlins: eine religionsphilosophische Untersuchung." Diss. Munich, 1954.

O'Brien, William Arctander. "Getting Blasted: Hölderlin's 'Wie wenn am Feiertage...'." *Modern Language Notes* 94(1) (1979): 569–586.

Ovid (Publius Ovidius Naso). *Metamorphoseon.* Ed. A. G. Lee. Cambridge: University Press, 1962.

Pellegrini, Alessandro. *Friedrich Hölderlin: Sein Bild in der Forschung.* Trans. Christoph Gaßner. Berlin: Walter de Gruyter, 1965.

Peters, Uwe Henrik. *Hölderlin: wider die These eines edlen Simulanten.* Reinbeck bei Hamburg: Rowohlt Verlag, 1982.

Peters, F. E., ed. *Greek Philosophical Terms.* New York: NYU Press, 1967.

Pongs, Herrmann. "Das ewige Herz in Hölderlins Dichtung." *Worte und Werte: Bruno Markwardt zum 60. Geburtstag.* Ed. Gustav Erdmann and Alfons Eichstaedt. Berlin: Walter de Gruyter, 1961.

Schmidt, Jochen. *Hölderlins später Widerruf in den Oden "Chiron," "Blödigkeit" und "Ganymed."* Tübingen: Max Niemeyer, 1978.

———. "Hölderlins idealistischer Dichtungsbegriff in der poetologischen Tradition des 18. Jahrhunderts." *Hölderlin-Jahrbuch* 22 (1980/81): 88–121.

Scimonello, Giovanni. "Hölderlin: Philosophie, Pantheismus, Tempor-alität." *Orbis Literarum* 28(1) (1973): 23–45.

Seifert, Albrecht. *Untersuchungen zu Hölderlins Pindar-Rezeption.* Munich: Wilhelm Fink, 1982. Münchner Germanistische Beiträge 32.

Stahl, Ernst Ludwig. "Hölderlin's Poetic Mission." *GLL* 2 (1948/49): 45–61.

Staiger, Emil. *Die Zeit als Einbildungskraft des Dichters.* 1939. 2nd. ed. Zurich and Leipzig: Max Niemeyer, 1953.

St. Augustine. *Confessions.* Trans. and Intro. R. S. Pine-Coffin. Harmondsworth, Middlesex: Penguin Books, 1986.

Stierle, Karlheinz. "The Language of 'Friedensfeier:' Festival in Revolutionary France and in Hölderlin." *Hegel and Hölderlin Conference.* Yale U (New Haven, CT). 9. October 1987. Rev. and exp. as "Die Friedensfeier: Sprache und Fest im revolutionärem und nachrevolutionärem Frankreich und bei Hölderlin." Haug 482–525.

Szondi, Peter. *Einführung in die literarische Hermeneutik.* Ed. Jean Bollack and Helen Stierlin. Frankfurt am Main: Insel, 1975.

———. *Hölderlin-Studien.* Frankfurt am Main: Insel, 1967.

Unger, Richard. *Hölderlin's Major Poetry: The Dialectics of Unity.* Bloomington, IN: Indiana UP, 1975.

Wahrig, Gerhardt, ed. *Deutsches Wörterbuch.* Gütersloh and Munich: Bertelsmann Lexikon Verlag, 1986.

West, M. L. "Excursus II: Time-Reckoning." *Works and Days.* By Hesiod. Oxford: Clarendon P, 1978.

Wilson, John R. "KAIROS as 'Due Measure.'" *Glotta* 58(3) (1980): 177–204.

Wohlfahrt, Günther. *Der Augenblick.* Freiburg, Munich: Karl Alber, 1982.

Wood, David. *The Deconstruction of Time.* Atlantic Highlands, NJ: Humanities P Int'l., 1989.

Index

—O—

—P—

—S—

—T—

—U—

—V—

—W—